T0302138

HOW NOT TO FAIL AT PROJECTS

STOPPING THE PROJECT MANAGEMENT INSANITY SPIRAL

HOW NOT TO FAIL AT PROJECTS

STOPPING THE PROJECT MANAGEMENT INSANITY SPIRAL

Claude H. Maley

CRC Press
Taylor & Francis Group

AN AUERBACH BOOK

Boca Raton and London

First edition published 2023
by CRC Press
6000 Broken Sound Parkway NW, Suite 300, Boca Raton, FL 33487-2742
and by CRC Press
4 Park Square, Milton Park, Abingdon, Oxon, OX14 4RN

Auerbach/CRC Press is an imprint of Taylor & Francis Group, LLC

Library of Congress Cataloging-in-Publication Data
A catalog record for this title has been requested.

ISBN: 978-1-032-74864-1 (hbk)
ISBN: 978-1-032-74450-6 (pbk)
ISBN: 978-1-003-47169-1 (ebk)
DOI: 10.1201/9781003471691

Typeset in Adobe Garamond Pro, Avenir LT Pro
by DerryField Publishing Services

Contents

List of Tables

Acknowledgments

Throughout my long professional career, I have met extraordinary people, from all walks of life, who have given me an exceptional exposure to working as a group to achieve common goals. The list of individuals is very long, and they all have allowed me to learn and expand my knowledge on the management of change by programs and projects.

I thank the many organizations which have given me the opportunity to practice my profession, especially certain individuals who have set me on my path and to whom I would like to express my sincere gratitude:

- Nick Dorling of IBM
- Freddy Ross of IMS International-Interdata
- Heribert Schmidt of Hewlett Packard Professional Services
- Tim Bailey of BP

I am greatly thankful to the consulting and education institutions Learning Tree, ESI International, Informa and Leoron for promoting and allowing me to present my vast collection of training topics to hundreds of organizations across all continents.

Special thanks to John Wyzalek of Taylor & Francis for his support and trust to complete my third book for his organization, and specifically to Susan Culligan of DerryField Publishing Services for her invaluable constructive feedback that significantly enriched the content of this book.

I extend my sincere appreciation to all my audience participants who have been present during my consulting and education assignments and who have provided me an extremely rich experience that has widened my knowledge and comprehension of what it takes to manage a group of individuals to achieve results.

My family's continuous support throughout the years has been immense, and it pleases me that my son Jeremy Maley is following in his father's footsteps in the world of project management.

About the Author

 **Claude H. Maley** is the founder of Mit Consultants and a Senior Management & Education Consultant, having started his career as a Systems Engineer with IBM, following reading Estate Manage-ment and Building Construction Engineering at the London School of Building. Claude has held senior positions with and for international organisations and companies in:

- Management
- Organisational management
- Programme and project management
- Sales and marketing.

His functional management and consulting experience has been with major corporations such as ABB, Alcatel, Areva, BP, Cadbury-Schweppes, Cartier, Caterpillar, Cisco, Ericsson, GE, Hewlett-Packard, IMS International, Motorola, Organon, Overseas Containers Limited, Pechiney, Renault Automobile, Sabic, Siemens, to name but a few. Claude also has extensive experience in the delivery of education and training sessions to major public and private entities in the Gulf countries.

Claude has broad experience in a variety of sectors which have exposed him to a variety of situations, forging a deep understanding of the issues governing management in organisations

- Engineering
- Oil and gas
- Production and manufacturing
- Pharmaceutical
- Distribution
- Transportation
- Marketing services

In his professional career, Claude has worked with more than 110 different nationalities in more than 80 nations on all continents and has:

- Held responsibility for a significant number of organisational initiatives and programmes/projects.
- Been engaged in change strategy definition, initiatives objective setting and programme/project implementation, strategic project portfolio management and instituting functional PMOs
- Delivered seminars, courses and training to over 15,000 participants across the continents
- Managed internal organisational missions and external commercial ventures
- Managed programmes/projects of duration from one month to five years
- Overseen budgets from $ 50k to $ 500m
- Participated in many assignments as leader of sub-systems and as an external advisor to lead managers

Claude is a professional speaker and instructor in topics ranging from general organisational, programme and project management to sales and marketing, leadership and motivation.

He is fluent in English, French, Spanish, and Italian.

Claude is author of the books *Enterprise Project Management,** *Project Management —Concepts, Methods & Techniques*† and his current book, *How Not to Fail at Project Management,* as well as educational courses and papers on business solutions, strategic and organisational goals, management of change by projects, organisational management and leadership. Claude is also a PMI® Certified PMP.

Claude may be reached at claude.maley@mitconsultants.com

---

* Maley, C (2023). *Enterprise Project Management.* Auerbach. ISBN 9781003424567.
† Maley, C. (2012). *Project Management Concepts, Methods and Techniques.* Routledge. ISBN 9781466502888.

Introduction

They say you learn from your mistakes. They also say that repeating the same thing and expecting a different result is insanity. However, in my many professional years I have bordered on insanity, and I note that many individuals in the position of Project Manager travel on that same path.

This short book is satirical, as I use humor to expose failures or shortcomings in everyday project life, with the objective of inspiring change.

I was first thrown into the project management arena totally unprepared for the professional path I was to pursue. Starting as a young trainee systems engineer at IBM®, my enthusiastic fervor was recognized by my superiors, who then proceeded to commit "professional homicide".

Let me explain. On a Thursday afternoon, standing by the coffee machine, I was approached by my manager, who said, "Hi Claude, I want you to handle

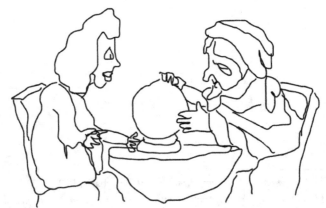

You will make the same Insane mistake again as before, and many times over!

something for us starting Monday". When I questioned him on what it was, he replied, "It's a small job of a couple of months". He informed me I would be a "team leader", whatever that was supposed to mean, and that this involved somewhat of a promotion. Never was the word "project" mentioned.

So there I was, elated that as a young man I was recognized in the company and that I had been given a "better" job. I spent the weekend celebrating with family and friends on my new career path. I never even considered what a "team leader" was or what the real meaning of a "small job" was.

Then it happened. On Monday morning I had a new job and zero experience to prepare me for what was to come. I was confronted with a new world of people I had not met before, a vocabulary of terms which made no sense to me, and my colleagues, who had been my work buddies the previous Friday, now saw me as their superior, expecting me to give directions and orders. Wow—how to start the week! I was the victim of a "professional homicide", as I was shifted from a position where I seemed to be successful to a new one for which I had no competence or skills and was destined for failure.

To all those who have had or are still having a similar experience, welcome to the club "How to Fail as a Project Manager". This book is a satirical commentary on the exciting and great profession of managing projects, and I propose what may seem simple actions to avoid failure.

I have held numerous positions in many enterprises, from supervisor to CEO, and for more than three decades I have delivered my consulting and education services globally in many industries in the field of Management of Change by Programs and Projects.

I consider myself a staunch promoter of the profession of project management. However, I have encountered a vast number of individuals in both private and public sectors who have no clue as to what the profession entails and commit "professional homicide".

As a reader of this book, I want you to enjoy this treatise on project management, as I have tried to make it provocative whilst being satiric, humorous and down to earth. My aim is to analyze the reasons for failure in project management with stories, anecdotes and sarcasm that highlight how organizations and project managers fall into an "insanity spiral". I do not go into the hard and soft skills that project managers must possess, as you can find this in my two previous books, *Project Management Concepts, Methods and Techniques*[*] and *Enterprise Project Management.*[†]

---

[*] Maley, C. (2012). *Project Management Concepts, Methods and Techniques.* Routledge. ISBN 9781466502888.

[†] Maley, C (2023). *Enterprise Project Management.* Auerbach. ISBN 9781003424567.

To all readers, whether you are practicing as a project manager, wanting to become one or having to work with and alongside one:

> I encourage you to instill Sanity Checks in the workplace and to become champions and promoters of this exciting project management profession.

I propose there are seven Sanity Checks to address:

1. The first sanity check is *how and when to appoint a project manager.*
2. The second sanity check is *the comprehension of why a project is needed.*
3. The third sanity check is *the understanding of the unknown.*
4. The fourth sanity check is *capturing who needs what.*
5. The fifth sanity check is *who does what.*
6. The sixth sanity check is *outside assistance.*
7. The seventh and most important sanity check is *engaging the efforts of others.*

I invite you to read each chapter in the sequence of the seven sanity checks.

- You will certainly find the first sanity check familiar to you, and this may well bring back memories with a smile on your face on the start of your professional experience in project management.
- The second and third sanity checks will guide you on how to overcome the misunderstanding that many have on the nature of projects and its management.
- The fourth and fifth sanity checks will surely strike very true in your constant pursuit to satisfy a host of individuals and at times your (sometimes seemingly) unsurmountable quest to secure resources for the project.
- The sixth sanity check is all about breaking the *we* and *them* syndrome when outsourcing in your project.
- The seventh sanity check is my favorite as it deals with people—the lifeblood of any organization.

Please enjoy the humor and satire I present in the sanity checks, as they are great remedies to the insanity often encountered in project management.

> Keep calm, and "may the force may be with you!"

Chapter 1

The First Sanity Check— How and When to Appoint a Project Manager

Moving into our first apartment, my wife and I decided to change the cupboards in the kitchen. We purchased three new cupboards to place above the kitchen sink area. Thinking I was a DIY guy, "No problems," I thought. So, with my "engineering" knowledge, I measured out the cupboards, prepared the pre-holes and mounted the cupboards, roughly aligned to the waterline. We filled the cupboards with crockery and glasses, and all looked good. "Job well done", I was satisfied to say to my inner self.

As we were in France, it was time for the lunchtime aperitif, and we went to relax in the lounge. Some five minutes later, the biggest crashing sound ever heard came from the kitchen. Rushing in to see what had happened, we saw that the three cupboards had been ripped out of the wall, and the crockery and glasses were broken all across the kitchen floor. My first reaction was to save our red and black goldfish, quivering and shaking, having been bombarded out of their broken glass bowl.

The moral of the story is simple: if you want to do a professional job, get someone who has the competence, skills and experience.

Would you give a 38-tonne truck to drive through the country pulling a refrigerated container for hundreds of kilometers to a person who does not know how to drive? Do not answer that.

Sure, you have no experience, you will learn on the job!

What amazes me is that every day, organizations give a project management task to individuals who have not had the training and experience to do the job professionally. The result is as bad as my kitchen cupboards.

1. Project Management Is a Profession

When organizations recognize that project management is a profession and not a part-time hobby, they will institute a training programme to prepare individuals for the job before assigning them to projects.

Education, training and exposure to projects for all personnel is a must, as organizations do not only perform operational tasks. There is a constant need for change to assess current processes and apply continuous improvements to maintain and sustain current operations, whilst management and those above determine new strategic avenues of growth whilst dealing with socioeconomic, geopolitical, environmental, health and safety, market competition and governing legislation issues that abound.

Those enterprises that can afford the luxury of having a department of full-time project managers may be able to address the changes required by the organization. However, even such a department will ultimately face a resource shortage. It is therefore incumbent on organizations that available personnel be assigned to manage a project as and when required—but only if the selected individuals are capable of handling the task.

2. Project Management Training for ALL

Organizations must institute a scheme for all personnel on the essence of project management and identify which individuals would have the desire and career perspective to eventually follow that path. This is an imperative, as organizations must react to and also be proactive in the management of change they all experience.

The project management scheme can be managed and/or supervised by the HR function and would initially make available assessment questionnaires to determine the knowledge gaps that exist within the employee pool. Those individuals who have declared an interest in pursuing their knowledge of project management will, in accordance with their line managers, follow fundamental project management training to be ready in case they are called upon to be assigned to "a small job"—a resonance of my first assignment described in the Introduction.

The project management scheme must also be extended to include more advanced features for established project managers, as the profession is perpetually evolving. Project management certification must be made available from institutions that offer these recognitions. All project managers must also have a well-defined career path that they can pursue in order to be retained as positive contributors to the organization.

Line managers and other functional managers would be scheduled to follow short-duration sessions on project management to allow them to fully understand and appreciate their roles.

Maintaining and growing knowledge and experience and enhancing competence and skills is an ongoing process, not a one-off. As Abraham Lincoln so wisely said, *"If you think education is expensive, try ignorance"*.

> **The conclusion is simple:**
>
> If an organization wants a project to be well done, it must provide its employees with the knowledge, exposure and experience to perform the job prior to being assigned to it.

3. Chapter 1 Summary

Putting the most appropriate person in the project manager position is of the utmost importance to realize the project's objectives. The project manager can come from the internal organization, by outsourcing or by hiring. There is no "best" person; whoever is appointed must have demonstrated the ability to deliver project(s) utilizing both hard and soft skills.

The most effective way to determine the ability of the future appointee is to conduct an assessment of the skills that person possesses.

The assessment will focus on the candidate's current competency, skill and experience in the following areas:

1. Organizational understanding
2. General project management aptitude
3. Project management knowledge areas (with reference to PMI®)
4. Personal communication, leadership and motivational skills

The Competency Assessment questions for each section can be adapted, expanded or even reduced, depending on the objectives of the organization.

Scoring for each question is on a 1 to 5 scale. Owing to the different nature of the questions and the way each question is formulated, the table below is a guide to the score to assign for the type of competency question:

Scoring	1	2	3	4	5
Ability	None	Low	Conversant	Proficient	Expert
Knowledge	None	Some	Average	Knowledgeable	Expert
Use	Not at all	Sporadically	Occasionally	Frequently	Always

Below is an extract of such an assessment:

Table 1.1 General Project Management Aptitude Assessment

Ability to	1	2	3	4	5
Implement and apply project management processes, best practices, and techniques within the framework of the organization's methodology(ies) throughout the project life cycle.					
Conduct ongoing analysis to identify and forecast budget and schedule variances.					
Develop and maintain a formal and comprehensive project plan that integrates and documents the project work—typically including the work breakdown structure, project deliverables/milestones, acceptance criteria, schedule, budget, communication, risk, and procurement plans.					

Chapter 2

The Second Sanity Check— Comprehension of Why a Project Is Needed

Every organization, be it private or public, strives to maintain and sustain its operation whilst it endeavors to grow and capitalize on opportunities. It faces constant forces and pressures from a multitude of areas, such as socioeconomic, geopolitical, environmental, legislative and market competition. Vision and mission statements abound in a large majority of organizations, and periodic formulations of strategic goals and operational objectives are the norm. Few, if any, are the organizations that can continue to operate without addressing changes in their environment—internal or external—as it operates in a dynamic world.

However, I find it amazing that many organizations manage change using an approach that is more *re*active than *pro*active. In addition, many fail to realize that change is inevitable in a future environment over which they have little influence. The insanity is drawing up change plans and believing in self-fulfilling prophecies to realize them.

Managing change is not easy at all, especially when you face a timeframe of many months or years. Complexity resulting from the dynamic nature of the organization's environment, associated with an extended timeframe needed for the change to be effective, will jeopardize whatever initial goals and objectives you may have had to achieve. Your change initiative will zigzag through the future, suffering multiple redirections and drawing upon excessive funds.

**My suggestion is
do Nothing and Hope for the Best**

However, you will still need an approach that allows you to structure how you would wish the change to happen, taking into consideration the various factors outside of your control. Sanity is using project management.

> Sanity is the need to recognize that change never happens as you would wish.

1. Projects as the Instruments of Change

The initial supposition is that the launch of projects is accepted as the principal approach to address strategic and operational changes, and that they create value for the organization.

Unfortunately, performance data collected on programs and projects reveals that, at a minimum, an astonishing 70% of them fail in all industries, reaching a staggering 85%+ in novel, innovative and technology domains. Failure is not meeting the organization's strategic and operational goals in any of three main vectors of scope contents, funding and timeframe, let alone quality or results. Failure subsequently regresses the organization's ability to operate in any way efficiently, taxes its financial investment capabilities and lowers the performance of the operational functional entities.

I am not referring just to pharaonic projects worldwide. Many are the cases of engineering and construction projects' vastly surpassing their original contents, budget and schedule. Run-of-the-mill internal organizational projects also fail due to, amongst other major reasons, the intricacies of cross-functional obstacles, differing and conflicting objectives and the lack of focus on project management. I also include those "small" projects that are supposed to last "a couple of weeks" but which turn out to grow exponentially in scope contents, plough through the calendar like a bullet train and generate soaring costs.

So—why is a project needed?

Let's start with two simple statements:

- Work that is repetitive and ongoing, employing staff hired to fill a job description and definition, to produce similar outputs on an hourly, daily, weekly or other basis against pre-determined targets, is *operational*.
- Work that employs staff with the competence and skills to perform tasks, produces a unique product or service funded over a planned timeframe, and comes to an end, is a *project*.

So, why do projects fail? Are the reasons the same as for operational failure?

Operational failures may be caused by the inadequate skills of employees, old or antiquated facilities, production process shortcomings, and other malfunctions. I will not expand on these causes, as there are many books and papers on the topic, and I will concentrate my discourse on projects.

2. Why Do Projects Fail?

Failure is basically the lack of success in meeting some previously stated and desired goal. In projects, the three major criteria for success are to deliver to scope, to budget and to schedule. However, I contend that the major failure of a project starts at its initial enunciation, as is explained later in this section. Suffice it to say that I have yet to experience success in delivering a project to its *initial* scope, budget and schedule. And I challenge anyone who has done so. Why? Because we live in a dynamic world, and nothing is static.

To comprehend why a project fails, let me start by reviewing how the need for the project is formulated from its inception and how it travels down several paths to become a worthwhile initiative for the organization to constitute and launch it.

The necessity to address an organization's strategic and operational change needs is a constant preoccupation for managers at all levels, as the measured operational performance data highlights those areas needing change. Change may also be forced upon the organization from external forces such as market or legislative changes. Furthermore, societal changes of all nature will also have an impact on the organization.

Individuals are tasked to analyze and formulate what and how to respond. Strategic planners, business analysts and the like, are assigned to assess the current situation and plan how to transit to a new and improved situation. This creates an overarching framework of change which will take the organization from one operational situation to another. Many documents are created; let us for now bundle these under a simple heading—the business case—and its development is in fact a project in itself.

Following its evaluation and approval, the business case is the principal entry point to the project. It sets for the project the objectives, scope contents, results to be produced, constraints, organizational readiness requirements, funding, resources and timeframe to be achieved, amongst other things.

Now, what is wrong with that? Seems perfect.

- First, the business case content's key goal is to state and illustrate the benefits to be realized by an extrapolation of all the costs to be engaged and the financial returns to be achieved within a desired timeframe. All of these are aspirational, as they include a vast number of assumptions, and they also enforce a project delivery deadline. I do not decry the work that all professionals perform to calculate and arrive at their numerical conclusions; however, I do decry the validity of their assumptions and the seeming ignorance that all of these will be challenged in a future that neither they, nor anyone, can control. As a result, the core business case will face a multitude of scope changes.

- Second, prior to any work being done by the project team, the funding to be allocated to the project is retro-calculated from either the break-even point or the payback period to fit the prospective benefits, thus imposing a financial constraint.

- Third, the project scope content is still described at a high level and is yet to be analyzed to determine what the project solution will be and if it will fit the required result expected.

- Fourth, the details and availability of the resources that will be required for the project—be they internal or external—are grossly defined and are still to be determined and identified.

- Fifth, the business case is competing for funding and resources over time with other ongoing business cases and projects and may be battling for priority, thus being imposed with a completely different timeframe than that originally envisaged.

- Sixth, and most important, without a comprehensive project portfolio management system that assembles all the ongoing and to-be launched projects, the organization has no way to appreciate whether it has the resource capacity and total funding to launch the project. Thus, projects are launched without any assurance that they will succeed in the perpetual justling of priorities and upper management decisions.

So, we can easily understand that projects are doomed to fail even before they are launched.

I will not list all the well-known reasons for the pitfalls that make projects fail during execution—they are too numerous, and you can easily find that list either online or in my book *Project Management Concepts, Methods and Techniques*. Suffice it to say that initiating a project "in vitro", without considering which other projects are ongoing, will lead to a continuous struggle within the organization to re-align goals and objectives, review expected business benefits, impose changes to the project's scope, re-assign resources, create cost overruns and dissatisfy all those who had high expectations.

Too many organizations perform too many projects concurrently, seemingly unaware that chaos is created, resulting in most, if not all, projects failing.

> To avoid projects' failing, an organization must institute a coordinated project portfolio management system, at both the strategic and operational levels, to launch the right project for the right reason, and expect to be nimble and agile to change in an environment of change.

3. Chapter 2 Summary

It has been stated many times that projects are the instruments of change. Project managers cannot work in a vacuum and only concentrate on delivering to scope, budget and schedule. A project's real success is measured in the organizational benefits it produces. Thus, the project manager must also possess an understanding of the organizational strategy to be achieved and the concurrent programs/projects that are ongoing and to be launched. The latter is best achieved by coordinating with a project portfolio management system, which must exist for the strategic needs of the organization to be realized.

As for the previous chapter, an assessment is to be conducted of the project manager's understanding of the strategic position of the project. Table 2.1 is an extract of such an assessment.

Table 2.1 Organizational Understanding Assessment

Ability to	1	2	3	4	5
Adhere to regulatory constraints, legal constraints, and all organizational policies and procedures in managing all project work					
Align the project to the company's organizational strategy					
Assess current business/industry and project management trends that may have an effect on the organization, its projects and its customers					
Assess interdependencies and other organizational processes and capabilities that need to be addressed during project implementation					

Chapter 3

The Third Sanity Check— Understanding of the Unknown

Without entering into a philosophical discussion about time, the past, the present and the future, I am amazed that many individuals do not recognize that any project will be performed "at a later date"—and that means the future. The most famous 20th century scientist is most probably Einstein, who once said *"Everything is determined by forces over which we have no control"*, which may mean that the future already exists, but we have not got there yet.

To get the gist of what he meant, I invite you to take a journey into his relativity theories. Suffice it to say for us mortals, the future awaits us. However, we may not know what the future holds, unless of course you have a crystal ball, tarot cards or any other artifact that can predict with accuracy what will happen next.

Projects will be performed in what most of us will accept is the future. We plan for the future, we anticipate things in the future, we forecast the future; we extrapolate on future demographic, socioeconomic and geopolitical futures; and every day markets attempt to calculate the evolution of share prices. None of these predictions become reality to a 100%. Many are, unfortunately, completely wrong, as the world has experienced since the beginning of this millennium: the internet crash of 2000, the financial melt-down of 2007–2008, the 2022 war in central Europe, and more recently the conflict in the Middle East, to cite just a few examples over the last 20 years.

And the reason is simple: the future is unknown, uncertain and unpredictable.

We make all sorts of assumptions and establish unrealistic aspirations. We pit optimists against pessimists, while the "don't-careists" seem impervious to any debate or discussion. Then we do our best to navigate the future to address and resolve our incorrect predictions as we endeavor to be nimble and manage change in an environment of change.

Humans have planned for projects ever since the earliest community needed to create something—either to evolve an existing thing or to build a new thing outright. From the creation or enhancement of hand-held tools to the construction of pyramids, humanity has been confronted with "how long will this take" and related concerns as to what resources, human or other, are available and whether they have the finances to do it at all.

Let me establish one premise from now on:

> Projects will be performed in the future in a
> world of uncertainty and unknowns.

You may disagree with that statement; however, it does not change the postulate that we really do not know for sure what awaits us "at a later date".

1. The Insanity of Fixed Plans

The incredible insanity in many organizations is the belief that any project plan is a self-fulfilling prophecy. From project managers to the highest levels of an enterprise, plans seem to be considered cast in stone, and everyone expects that the contents are static and will actually happen. I have never delivered any project according to its original plan. And I am very proud of that, because the project contents were full of incomplete information, miscomprehensions, unknowns and assumptions, the near impossibility to arrive at details, optimistic estimations and predictions, no full understand of the working environment and the associated risks, management pressure, and on top of all that, and more, just wishful thinking.

Unfortunately, higher management often suffers from what I call the "Grandma Syndrome". Let me explain.

Proud parents will make photos of their offspring available to grandparents. Forget about internet and digital photos, grandparents want real photos to frame and put on the mantelpiece or hang on the wall so they can be seen all the time.

When the time comes that the grandchild visits—and that may not be every month—the first thing Grandma says is, "My, you have grown", to which we

**So, you're telling me you know what you know,
but you don't know what you don't know!!**

may reply "Yep—we feed them". Obviously, Grandma compares the real child to the photo, and there are many differences, especially in body size.

What does top management do? Because of their extremely full schedules, they revisit projects on their institutionalized quarterly meetings, if that. And what do they have to compare? The mantelpiece photo of the original project plan to the actual status of the project. And it is no surprise that they utter the famous Grandma Syndrome line, "My, that project has grown—and changed!" You hesitate to respond with an insulting quip, as you deem it pointless unless you have a new job waiting for you elsewhere.

The insanity is that so many individuals just do not understand that plans are made to be changed, as their contents are based on incomplete information, are full of assumptions, speculative risk management and assessment, and that all the estimates of effort and duration can only ever be approximately right.

1.1 Sane Project Planning

Sanity in project planning is accepting that you are predicting the future and that you will be wrong. Irrespective of the future time window you have envisaged—be it short-term, for the next week, or long-term, for the next decade—your speculations will depend on the information you have on hand and any experience and exposure you or others have had on similar projects.

In the above section "Why Do Projects Fail", I explained that the business case and other enabling documents describing the scope brief that are made available at the start of a project are all aspirational, as they state what is known of a current situation and what should/could/may be a future situation. A problem is given, if it is completely described in detail, and a somewhat overarching solution is proposed, which may not have any resemblance to the organizational, technical, environmental or any other reality that awaits the project.

As I have repeated more than once, the future is uncertain and unknown. And what we do not know about the contents of the business case and what awaits us once we launch the project is of major concern.

1.1 Socrates and Wisdom

It is said that Socrates enounced words to the effect that, *"Wisdom comes when I know that I don't know"*. We will honor the man by establishing at the outset of any project an inventory of all Knowns. This is achieved by reading through all the documentation made available to the project and recording specifically precise statements. Do *not* presume or assume. When in doubt, when you fail to understand a given statement because it is ambiguous, undefined and/or esoteric, put this to the side to be classified in an inventory of Unknowns.

Seek common agreement with all identified stakeholders and those organizational entities identified in the Knowns inventory to ensure that everyone is on the same page on the given facts. For example, if the project brief states that the manufacturing/assembly process will be enhanced as a result of the project and will affect the current 85 employees, ensure that that is a fact—there are 85 employees currently on manufacturing/assembly. You can imagine all the other examples I can give you. The important thing is that the Knowns are known to everyone and constitute the facts of the project at the time of their identification. You can read between the lines that any subsequent changes to these facts will affect the project scope.

Then, to be a good disciple of Socrates, you tackle the Unknowns inventory. That could well be a very long list of items. To establish as comprehensive a list as possible, you proceed by separating the Unknowns into three categories: Missing, Ambiguous, Incomprehensible.

- The **Missing** category includes all the information that is just not present in the documentation at hand. You use your experience and that of other project team members, and you can assist the categorization using a checklist, which will ease the discovery of missing Unknowns, as it will give a simple Yes or No to the existence of the required information—for example, from simply checking that the coordinates of an identified stakeholder are given, to identifying missing definitions and descriptions of current ways of operating, such as how the manufacturing/assembly process is currently defective. There are umpteen examples possible, and you can tailor the checklist to your environment. Pilots on jumbo jets who have flown hundreds of hours always go through a checklist before starting engines, else the flight may be of a short duration.

- The **Ambiguous** category includes all definitions and descriptions that contain un-defined or ill-defined vocabulary, such as adjectives and adverbs. What does a "user-friendly system" mean, or a "rapid response time". Well, nothing. The one that glares out as being totally ambiguous is the famous phrase "the interface description will be defined at a later stage", as well as the classic three-letter acronym "TBD"—To Be Defined. So, all these statements enter the Unknowns inventory.

- Lastly, the **Incomprehensible** category includes all the parts of the documentation available that are just that: incomprehensible. This includes the abundant use of acronyms (if you ever encounter the acronym TLA, this means "Three Letter Acronym"). Apparently, the longest known acronym, according to the *Guinness Book of World Records,* is 56 letters (54 in Cyrillic) in the *Concise Dictionary of Soviet Terminology, Institutions and Abbreviations* (1969).* Check it out if you want to break the record.

The majority of Incomprehensible Unknowns comes from esoteric terminology, where you seem to be placed in another universe, as scientific, technical and other "expert" terminology totally escapes your comprehension.

Then comes the grammatical compositions. Some 5% of the world's population are part of the Anglosphere (countries where English is the main native

---

* https://doi.org/10.2307/2493532

language). Anglophones—speakers of English—may rise to 25% of the world's population. Since many projects may be in English, this leaves extraordinary liberty of writing and interpretation, as an extremely small proportion of individuals can be classified as literary experts.

1.2 Addressing Your Unknowns

You now have as comprehensive an inventory of Unknowns as you can identify.

The next step is to change Unknowns to Knowns. I propose you use what I have coined as the GAP approach (okay, an acronym). It stands for Get, Assume and Park. And you proceed down the inventory, Unknown by Unknown.

Get the information

From the classified Unknowns, you go in search of the Missing, Ambiguous and Incomprehensible items. This will involve meeting with individuals associated with the project who can provide information and facts on these items. This can be quick, such as, "How many employees need to be trained on the new system?", and you get the answer 150. Similarly, you can get a comprehensive reply to the question, *"What do you mean by rapid response time; can you quantify that?",* and get 125ms. For another question like, *"In a construction project, can you explain what a building envelope is?",* you may get a longer reply, which you may eventually understand and turn into a fact.

At times, project effort is required if the missing item is to collect and analyze all the stakeholders involved. It may necessitate analysis of the detailed project requirements with these same stakeholders, which will then constitute the engineering phase of the project and thus need team members with the competence and skills to develop the said document. Thus, to establish the facts, some missing items may be collected immediately, whilst others will need time and effort to be established.

All these newly acquired facts reduce the Unknown inventory and are now collectively part of the project scope specification. Anticipate that you will not be able to answer all missing items.

Assume

Assumptions are part of life, and therefore of projects. Let's not break down that word, as many people do. Assumptions are made when a precise answer cannot be given. To follow the example above to the question, *"How many employees need to be trained on the new system?",* it may be that a precise reply cannot be given when the organization may need to either hire new personnel or downsize staff. This leads to an assumption being made of the number of people to train, within a range, and all parties agree to make a common assumption that can be validated. For example, the answer could be, *"We anticipate approximately 150 people, with a maximum of 175"*.

When all parties agree on the assumption, the Unknown becomes an Unknown/Known and enters the Virtual facts specification—Virtual, because even though it has been agreed upon, it could still be subject to change. However, as an agreed and validated assumption, you will use the virtual fact along with the real ones, as both now constitute the core project scope specification—you will plan to train a maximum of 175 people. Every assumption that is agreed upon and is validated reduces the inventory of Unknowns.

Parking the Unknown

Parking occurs often in meetings when during a discussion on a specific topic, all parties conclude that there is not enough information presently on hand. Parking in this instance will let us put aside getting a concrete answer to this topic until we can get more information.

Parking will also be invoked when it is not possible to make a valid assumption because more analysis would be needed. Parking therefore means that the Unknown remains Unknown. What this means is that whatever is Unknown can either be excluded from the project, descoped, or placed in a temporary state of limbo, waiting to be addressed.

Let us continue with the example, *"How many employees need to be trained on the new system?"* If nobody can provide a number, because they just do not know if training may even be required, then the Unknown is dropped from the inventory and becomes a Definite Out of Scope item for the project. This is recorded, such that training is no longer in the project scope. However, if training is going to be required, the number of employees may depend on the result of a new organizational structure and hiring plan.

And if this plan will be formulated by management following the yearly budget negotiations later in the year, then we have a conundrum. Is training in or out of the scope? If training is sure to be included in the project scope, then this Unknown will remain in limbo until such time as the number of people to train is known. The inventory item now becomes a Known/Unknown because we expect to get an answer, but at a future date.

In this way, we prepare and anticipate that a managed change request will apply to the training project scope contents once we are provided with a number. All Unknowns that are placed in limbo awaiting a future management decision become Known/Unknowns and constitute a list of to-be-anticipated change requests, and management has to improve this list.

> Remember—the only moment you will have collected all
> Knowns in the project is at its conclusion.

1.3 Doing Your Homework

Project planning should not be depreciated as just an administrative task, considering the true nature of the intensive intellectual and engineering effort engaged by project participants. A plan is a detailed proposal of a solution for doing or accomplishing something.

Engineering is the application of science and technology for the specification and design of a conceptual paper solution that enables others to materialize the solution as a building, a construction, a system or a process. And we all accept that as being noble.

As a project manager, you and your team will engineer the solution of your project by developing the detailed project scope and the detailed individual product scopes pertaining to each deliverable by doing the following:

- Determining the work activities required, the nature and types of resources required and the durations of these activities
- Establishing the funding needed for the project's defined resources
- Analyzing and assessing project risks and instituting processes for managing change requests
- Communicating with stakeholders
- Ensuring quality and conducting procurements

And we put all those actions under "project planning"—I find that debasing. I am open to debate this.

Doing the project homework is to engineer the eventual contents of the Project Plan—yes, I will use that designation for now; I will change the world later since I am busy this week.

When you engineer a new machine or piece of equipment, using modern CAD/CAM software, you can turn it around in all positions, even test it by simulation. However, here you are dealing with solids, following physical and material behaviors and characteristics, and you can test the stress levels and where they can break—all before you even build and materialize the thing. That is a great achievement.

A project plan, by definition, is a plan for the future. However detailed a plan can be, there are still unknowns and assumptions that are embedded. And you are dealing with estimations and the hope to secure whatever resources you need some time in the future. You may not be able to arrive at the lowest detail unless you spend your time in breaking down the plan to the smallest screw and bolt, as an engineer with a machine can and must do.

I will not itemize all the actions you have to perform to produce a project, as I invite you to find this in my book *Project Management Concepts, Tools and Techniques*. In summary, your planning will consist of the following actions:

- Develop the product scope for each deliverable to produce.
- Develop the project scope of all the activities to perform to produce the deliverables.
- Analyze each activity to determine the resources required to complete it and to develop an estimate of duration and cost for each piece of work. Concentrate on being approximately right as you consider abstract resources to perform work in the future, and do not contemplate a stable and static future.
- Structure the sequence of work by producing a network of activities.
- Insert the results of the risk assessment and responses.
- Add the procurement and lead times activities.
- Complete the plan with the quality assurance activities.
- Establish a preliminary schedule and budget.
- Institute a robust change request management process, as you are not naïve enough to believe that change will not happen—and from all sides.
- Develop a tight communication plan for all stakeholders and indicate their contribution.

Projects in execution will navigate like a small sailboat in a large, choppy ocean. They will never reach their destination in a straight line. Every sailor or adventurer knows that, and they prepare for months before setting out to sea. And guess what? Some never make it to the finish line.

A project plan is as good as its contents. Remember—you plan to replan, but:

> If you fail to plan, you plan to fail.

2. Chapter 3 Summary

Projects create change, which means that the *present* situation will evolve to be different in a *future* situation. The project boundary—the scope—can assist in determining the present situation within its perimeter by conducting a situation analysis. But that does not include all other present situations outside of its periphery unknown to the project manager.

As for the future situation, only an imagined picture can be drawn, and it will depend on how far in the future this picture extends. Regardless of the extent of the future time window, uncertainty, and therefore unknowns, exist.

Uncertainty is a representation of the possible range of values associated with either a future outcome or the lack of knowledge of an existing state.

Uncertainty increases with respect to the anticipated project duration, across a spectrum of terms over time: immediate, short, long and far. Uncertainty may be low and manageable in the immediate term, growing to be high and speculative as the far term is reached.

This chapter has covered how a project manager can proceed to tackle knowns and unknowns and the challenges to project planning and realization. You are invited to refer to the chapter on Project Risk Management in my book *Project Management Concepts, Methods and Techniques*, cited above, for this topic.

However, the most important approach and attitude to managing uncertainty in projects is to be flexible and ever-changing, reacting to evolving situations and accepting that the originally targeted scope, budget and schedule are no longer valid.

Chapter 4

The Fourth Sanity Check— Capturing Who Needs What

Projects in organizations produce results that are for others to exploit, and there are numerous "others". One of the insanities is that not only is it unknown who all the others are at the start of a project, but those others identified at the beginning may no longer be around during and at the end of the project.

These others all come under the heading of "Stakeholders".

The interesting thing about stakeholders is that the word originates from the word "stake"—to denote a stick or a post sharpened to be driven into the ground. The stake could then be used to delimit a parcel of land around which a fence could be erected. Land partitioning, as every surveyor will tell you when drawing up a cadastral plan, is where a land area is accurately measured and is then recorded to delimit the internal and external boundaries of a property. A cadaster is then produced for each parcel as an official register of the quantity, value, and ownership of real estate used in apportioning taxes. Holding a stake, therefore, is laying claim to an area.

Furthermore, placing stakes in an area and claiming that parcel of land can be controversial, as it implies taking over control of another person's property or land. Moreover, stakes delimiting the perimeter of a parcel of land clearly intend to separate that area from any other pretender to that same space—there is no overlap. In other contexts, a stake is also used for many things, such as a betting stake, burning at the stake, and we cannot forget plunging a stake into a vampire's heart.

Returning to a project, we can envision that the change it creates will cover and affect a certain area of the enterprise—just imagine casting a net over the company's org chart, where the fish are the functional groups and the employees within them. Since organizations are commonly structured as a top-down pyramid from the CEO down to the operational level, it is important to determine the extent of the boundaries that the project encompasses and which organizational entities and employees are within that space.

This is where project stakeholders come into the frame. The only claim that project stakeholders hold are the organizational areas for which they have responsibility. However, they may not be the sole holders of that claim, as higher-level management and cross-functional organizational processes may well overlap on what initially seemed a unique ownership. This will lead to many different stakeholders enclosed within one or more organizational areas.

4.1 Stakeholders Have Needs to Fulfil

Returning to the first statement—that projects produce results that are for others to exploit—when areas overlap, it becomes complicated to determine which organizational entity has the overarching, final say as what results to produce and their specification for the project, creating *en passant* frictions and potential conflicts of interest at the managerial responsibility levels.

Meanwhile, all who are part of the stakeholder domain—namely, operational employees—are the ultimate users and exploiters of the project's results. These individuals are automatically stakeholders. It remains to be seen how much influence they may have in describing the results that are deemed to be operationally beneficial for their daily use.

In addition, all individuals who perform any work activity on the project, whether they are sourced internally or externally to the organization, are stakeholders too.

You, as a project manager, now face a conundrum—who are all the project stakeholders? A pool of individuals and entities is to be identified: where do they reside, internally or externally; what are their needs and specifications of the results to produce and where and from whom to capture these; which stakeholders are directly or indirectly affected by the project; which stakeholders are favorable to the project, and which ones are opposed—not ignoring those stakeholders who show no interest but are still part of the project stakeholder pool.

The issue becomes more complicated, and at times complex, as there exists a management hierarchy. We can discuss at length the virtues of flat organizations;

however, for a large majority of enterprises, the pyramid seems to be the preferred structure. That creates, from the bottom up, positions for supervisors, managers, directors, VPs, SVPs right to the top. That hierarchy of managers places a heavy burden on the organization, where countless hours are devoted to the supervision of others. So, how wide has the net been cast to capture all stakeholders?

You need to determine which stakeholders are key—where "key" may mean important, influential, powerful, etc. Many of the key stakeholders will be obvious, as they would be mentioned and identified in enabling documents such as business cases. The sponsorship group members are also your key stakeholders. I take a "3D" approach to identify stakeholders:

- **Drivers:** From these key individuals or entities, you can now categorize those who are the *Drivers* of the project. They are for and behind the endeavor, as they want the results to fulfil their strategic and tactical goals and objectives. The Drivers will be the major providers of the project scope contents and the ultimate decision-makers as to any evolution of that scope.

- **Deliverers:** The Drivers will be accompanied by the *Deliverers*. These cover the individuals and entities who carry the responsibility for enabling the organization to achieve the business goals and benefits. The Deliverers will for the most part be from the functional operational departments and will drive the scope contents, with a major accent on the operational environment changes to be made. In certain cases, an individual or entity may have the dual role of Driver and Deliverer.

- **Doers:** A third category of stakeholders will be the *Doers*. This category covers all those who will perform activities and tasks for the project. Doers will be from the internal organization and from external providers and public agencies. These stakeholders require clear work specifications and well-defined descriptions of the deliverables to be produced.

As a guiding principle, stakeholders' needs drive project scope contents.

> It is therefore imperative that you institute and follow a robust project stakeholder management process which casts a widespread net across the organization and the concerned external entities.

4.2 Who Makes Decisions as to What Is Needed

Humans are wonderful creatures. You can be sure that what is said one day will be changed another day. We have demonstrated over millennia our capacity to change and evolve according to the conditions that surround us, else we would have been extinct a long time ago. Many individuals do not change easily or just cannot or do not want to for a variety of reasons ranging from comfort zones to a fear of the future and the unknown. Stubbornness in maintaining the status quo and/or insisting on remaining where they stand or upholding their beliefs will at all times create barriers to those who believe that the change is necessary and beneficial.

I will not expand on the resistance to change here, as there are many books and articles on the subject. You may also want to read about it in my book *Enterprise Project Management*.

I *will* expand on the decision-making process that you as a project manager will need to comprehend to be conversant in the matter.

In any social group—and project stakeholders are such a group, be it temporarily and sporadically—order and structure have to be the norm. And decisions taken would follow a hierarchical flow. Rare are the organizations where an assembly-line employee, or any employee for that matter, needing to change a tool can just go out and buy a replacement and spend whatever amount of money without a higher authority approving that purchase.

I was privy once to the Expenditure Authority Level manual of a major organization, which detailed down to each position the limit and extent of the purchasing authority and the approval lines that had to be followed. The manual was 137 pages thick. In another instance, I was told by a senior engineer and draughtsman that many years ago they could not change their pencils unless they were used down to 10cm, as that is the minimum size to be able to hold one and still draw. Obviously, I am talking about an event that occurred many years ago. However, the point is that decision-making authority levels will often hamper the flow of the execution of a project.

So, who makes decisions in projects? The naïve answer is the project manager.

Before attempting to answer the question, let us review what decisions need to be taken; then we can get a better feel for the order of things.

4.2.1 Management Layers in Decision-Making

The first decision is the one taken to launch the project of change to satisfy and achieve an organizational goal. This decision is made by those who would become the Drivers (stakeholders), as they would be the promoters for the project to be realized. However, that decision may need authorization for the funding and would need others (who may well not be part of the project stakeholder pool) to validate the investment.

Other individuals, also totally unassociated to the project, such as management and/or steering committees, will have their say as to the soundness of the proposed project, as it may impact and be contrary to other strategic initiatives in progress or to be launched. As decisions get pushed upwards to higher levels of management in a pyramidal hierarchy, upper management is the furthest from the project realities. They would have what I would call a "satellite view" of the project and not a "helicopter view". Those who may be in a position to argue the decision as unworkable may miserably fail, as the decision-makers' positions and associated power may not allow any disagreement. Finally, upper management may also hold final decision-making authority if and when there are disruptions in the organization that may invalidate or even stop the project.

Therefore, the Drivers may not be the major decision-makers.

Decisions Made by the Sponsorship Group

Within the confines of the project, recognizing that it has gone through the ring-of-fire approval process and that upper-level management intervention is always a possibility, decisions made during project development will and should be escalated to the sponsorship group. As an assigned project manager, your success parameters have been defined in the classic way: deliver to scope, to budget and to schedule according to an approved baseline.

When any of these parameters are outside the margins that may have been set, the sponsorship group will have the final decision as to the actions to undertake. You will certainly have presented the status of the project and the reasons for deviations, as well as propositions as to how to steer back, if possible, the development to a new and acceptable goal. You may be able to argue for your plan, but the sponsorship group will have the final say. In cases when that is not possible, escalation takes another step upwards, and we find ourselves again in the pyramidal decision-making structure.

Let us ponder for a moment what this means. A multi-level management structure leads to additional time for considering acceptable and adaptable decisions, enduring approval layers, all resulting in slower responses. The consequences are evident—the project stops and waits for a decision. And whatever plan you presented to the sponsorship group to overcome the project variances is now null and void. *C'est la vie.*

Decisions Made by the Project Manager

There are of course decisions that you will be able to make as a project manager without having to escalate to the sponsorship group. These will often cover issues in technical and human resource allocation matters, and you can exercise your decision-making authority—which should be clearly stated in the project charter—when the project is within the boundaries of the given margins of variances. Any funding issue requiring a decision would most probably be escalated.

However, you would likely not possess the faculties or knowledge to address all the highly technical issues. In addition, as you do not want to slip into a dictatorial or authoritarian role, you must make decisions with the assistance of and input from your core team members. You must also seek advice from subject matter experts, who, as their definition makes clear, will have the know-how to propose adaptable solutions. Your decision-making role takes on a consensus style, involving as many project members as necessary. This approach must also be considered when work is performed with outsourcing providers.

Dealing with resource allocation issues is a different kettle of fish. You can address these if and when you have a large and polyvalent team and they all report directly to you. However, as is very often the case, many project performers (the Doers) will be sourced from internal functional departments, opening the door to complicated matrix-management problems. This will lead you to have to negotiate the assignment of human resources with managers who are part of the stakeholder pool. These managers, in their enthusiasm to demonstrate their authority, may often hamper and slow down decision-making for releasing resources, rather than speed up your assignment request.

The consequences are evident—delays in the execution of the concerned project tasks. Do not fall into the trap of accepting just any human resource assigned by the functional manager by failing to verify whether their competence, skills and availability are equal to or compatible with the required profile for the project.

The conclusion is clear—decision-making in projects can call upon a potpourri of individuals, sticking their fingers into the problem that needs to be resolved. The more an organization is vertical and pyramidal in structure, the more layers of decision-making are traversed. And that is the best way for a project to fail, as it will become materially impossible to fulfill the scope within any set budget or timeframe.

> Organizations must establish clear stakeholder roles, responsibilities and accountability; eliminate wasting time in decision-making; and allow project managers more autonomy.

5. Chapter 4 Summary

Capturing the needs, requirements and contents of the project scope is a continuous effort for the project manager. No project scope is static, owing to the potential inconsistencies of its original contents and the ever-changing stakeholder constituent group and their individuals' evolving needs in a dynamic world.

Stakeholders will change their needs because of factors that affect their sphere of responsibility. The project manager must consider and prompt analysis of the different project stakeholders and their influences on the scope throughout the project lifecycle, then capture the evolution of needs. Stakeholders have concerns, and these are influenced by their respective focus on matters such as these:

- Attitudes to work and leisure
- Business cycles
- Competition law
- Consumerism
- Disposable income
- Employment law
- Energy consumption
- Environmental impact
- Environmental legislation
- GNP trends
- Health and safety law
- Income distribution
- Inflation
- Interest rates
- International/European agreement/law
- IS/IT developments
- Levels of education
- Lifestyle changes
- Local government/devolved administrations
- New discoveries
- Population demographics
- Rates of obsolescence
- Regional legislation
- Social mobility
- Speed of technology transfer
- Taxation policy
- Unemployment
- Waste disposal

This chapter has covered how stakeholders can be classified and the challenges in the collection and evolution of their needs and requirements.

The project manager must proceed with regular and frequent stakeholder analysis to determine how the scope of the project is evolving and to assess the changes that are to be made.

Chapter 5

The Fifth Sanity Check— Who Does What

We all know of Superman and Superwoman. They have extraordinary powers and can perform a multitude of things single-handedly.

You, as a project manager, are not a super-hero.

It is materially impossible for you to perform every work activity and/or to possess all the competence and skills required for the job. You must be proficient in managing the project and not dedicate time to the performance of work tasks, unless there is a resource shortage, which will lead you with less material time available to fulfil your prime responsibilities.

Projects need team members to perform work and internal and external organizational entities to provide those human resources.

Your first challenge is to be supported by a core project team. This will consist of a small number of individuals who will work closely with you and be assigned major scope sections or partitions of the project, which would usually revolve around the project deliverables.

The project can also be partitioned by functional area, discipline, geography or other sensible way to break it down at this initial stage. The assignment of core team members is a very important step that has to be done at the project charter level. Otherwise, you will be left alone to structure the project meaningfully, and you may not have the necessary knowledge or experience to do so. Certain individuals will be assigned permanently to the core team, as they will subsequently be project leads in their respective areas.

**Is it too late to give you the
Project's Risk Assessment?**

You and the core team develop the high-level breakdown structure of the project and how the product scope(s) and associated deliverables will be distributed. The product scopes relate to the project deliverables, and subject matter experts should be assigned to perform that work. These individuals, possessing the required competence and skills, will analyze and engineer how each deliverable will be designed, fabricated and produced in the project implementation phase. The product scopes are collated to constitute the product requirements specifications, to be designed later as the engineered solution.

The kernel of the project scope will consist of determining all the activities, including design reviews, that need to be performed to realize the product specifications. These activities also encompass those related to the deliverables, such as procurement, quality assurance, testing and commissioning. Other activities are to be identified to address the transition to the new solution: organizational readiness, user training, user documentation, support contracts as required and other health and safety legislative certifications and approvals where necessary.

The project scope activities are further expanded by the results of the risk assessment and response plan.

Project planning is conducted in parallel with the development of the product scopes, and schedules and costs are progressively elaborated from a high level to as low a level as is possible or required.

All seems good on paper and requires people to perform the work. And now comes the huge difficulty—where are these people?

5.1 The "Hunt" for Project Human Resources

Human resources for any project are scarce and rarely abundant in the profiles required. Insanity leads to decision-makers' assuming that human resources with the pre-requisite skills and availability will be available in the future. You will certainly be the first named person on the project—that is a given. All other individuals needed to perform on the project can only be provided by internal functional departments and from external human resource sources. People are required during the planning phase and in the execution/implementation phase, and they obviously have to possess the competence and the availability to do the work.

Here starts the hunt for project team members—I use the term *team member* for any person who will perform on project activities, be it temporarily or on a permanent assignment basis. And from now, when I write *resource(s)*, this refers to people, individuals, persons, staff, etc. As you contemplate the product brief, you must first realize that you have no idea where resources will come from and whether they will have the competence or availability to do the work. You are now in the realm of resource management, where competence and availability are the cornerstones for any assignment.

The first thing you must determine is the required profile of this (for now) anonymous resource: skills, competence, experience, availability. After having drafted a job description, your actions from this point onwards are similar to the hiring process:

- Resources can be provided by the internal functional departments. However, the functional departments' primary roles are to achieve their operational goals. That is why they have resources in their structure. Availability of resources, along with their competence matched to the ones needed for the project, have to be assessed.

- Those organizations that function with a program/project department are also able to provide resources; however, the problem remains the same as for functional departments.

- External providers may be able to release resources on a time-and-materials basis, but here again the problem remains as above. You may opt to contract out packages of work to these providers, thus shifting the onus onto them. The problem does not disappear; it is now located elsewhere, and the providers face the same issues with resource management.

As stated above, the first resources are for the core team. You initially may have to break down the project alone in sub-sets to determine the profiles of core team members. A point to bear in mind is that business analysis may still be required, depending on the contents of the business case, especially when there will certainly be incomplete information—for example, the full list of stakeholders and their needs—and it is reasonable to assign a competent person to enhance and complete the business case.

You must secure these resources at the same time that the project charter is drafted and approved. The sponsorship group must make this happen, else you will be left high and dry, having to conjure up how you will even understand what to do and how to do it. You must impose your choices for the project to have a minimum chance of not failing.

Once core team members are secured and aligned to the job descriptions you have established, initiation and planning phases can commence. This may well be easier said than done, as core team members, who are also not superheroes, would need resources to perform their sub-project work of developing the respective project deliverable requirements and specifications.

Now begins a second round of the hunt for resources. You and the core team members, following an initial high-level breakdown of the project and its subsets, must assign resources to perform the engineering and planning activities.

5.2 The Challenge of Matrix Resource Management

A matrix, in mathematical terms, is a rectangular array or table arranged in rows and columns. In organizations, this is illustrated in the top-down pyramidal org chart, in which business units and functional and operational departments determine the number of rows in the different columns. Management experts will propose that the silo effect in organizations promotes for each column in the enterprise to foster specialization and efficiency. Difficult to argue against that.

However, the cross-functional communication and information flow needs of any organization will at times be confronted with insurmountable obstacles, as each silo may protect its territory.

The insanity for projects lies when different resources are required from different parts of the organization, yet management assumes that all cross-functional obstacles can be resolved satisfactorily, since all departments are working for the better good of the enterprise. Take a moment to smile.

Resources will be provided by internal or external organizations. Resources provided by external organizations are discussed in Chapter 6, The Sixth Sanity—Outside Assistance, and you can jump there when you wish.

For most projects, unless completely sub-contracted out, you are faced with securing resources initially from the internal organization to address product scope work and project scope specification and planning activities. I go into more detail on resource assignments for the project execution/implementation phase below.

5.2.1 Securing Internal Resources for the Product Scope

You are now in the classic framework of matrix management. Internal resources are there for a specific purpose—they have been hired to perform operational work, and their direct line managers are the only ones responsible for their assignments and priorities.

A major stumbling block during the planning phase is to negotiate with functional managers as to the availability of resources that are subject matter experts for developing the product scope specification and design. These managers should/would have been identified in the stakeholder analysis conducted at the initiation of the project.

So, you could assume they are on board to facilitate the resource assignment. You would be naïve to assume that you will easily secure the resources unless there is a mandate from higher-level management. You may have to do with whatever resources are available for the project, in conjunction with their operational workload.

If and when you do secure the resource, you need to ascertain whether they possess the necessary profile. Here again, you may have to do with a below-par profile, but you may even be lucky and secure an above-par profile. The consequence of internal resource assignments that do not match both profile and availability will stretch the original schedule for the planning phase and most probably lower the quality of the product scope output.

In addition, the assigned resource will have two "bosses" to report to during the assignment. You can expect that when there is an increased operational workload, the line manager's priorities will outweigh those of your project, and the resource would be pulled back, resulting in more schedule slippage—and you are still in the planning phase.

As project planning is principally a product-scope requirements and design/engineering phase, you must expect that the analytical work done and the details reached will most probably challenge the original scope contents as described in the business case. Refer to Chapter 3, The Third Sanity Check—Understanding of the Unknown, for more on that. This will extend the original duration of the resource assignment and may even call upon additional expertise to be allocated to the project.

5.2.2 Securing Internal Resources for the Project Scope Planning

In parallel to the product scope work, which requires profiles of subject matter experts, you and the core team must now concentrate on building the project plan constituents. This is classic in determining the activities to be performed in the execution/implementation phase: identifying resource needs and estimating duration and cost for each activity, building a schedule along with a resource loading chart and cumulative cost curve for the project, and ensuring that risk responses are translated to additional activities to be included in the project.

There is nothing wrong with that sequence of events, which would be iterative as you progress during the planning phase. The challenge is whether you and the core team members have the availability and competence to actually develop the contents of the project plan. You would need a profile that extends to expertise in estimating, costing, scheduling and the use of modern project-management software. If you do have this expertise, it would be possible for you to develop the detailed project plan; however, once again this will cut into your available time for managing the project in its entirety.

The more rational option is to call upon subject matter experts in the areas stated above. That would involve securing internal resources to perform that work, and they may not be required on a full-time assignment basis.

If your organization has a program/project-dedicated department, then it may be that the experts are part of that entity and may be assigned to the project. Here again, availability could be a stumbling block, or else you may have to share the resource with other ongoing projects.

> A lack of management support to allocate and assign resources during the planning phase will undoubtedly delay the project effort, even before the start of the execution/implementation phase.

5.3 Challenges for Project Execution/Implementation

A project plan principally contains all the work to be performed to produce the deliverables, based upon the assumed availability and profiles of resources to be assigned and allocated in the future. *[Wishful thinking is another demonstration of insanity in projects.]*

This plan becomes the execution/implementation baseline once it is approved by the sponsorship group and key stakeholders. However, to arrive at this approval, multiple iterations are needed to determine if the major constraints of the initial scope, initial budget and initial schedule are achievable. It will be no surprise to know that from the initiation of the project to the baseline approval step, changes will have been made to the initial scope, affecting the initial budget and initial schedule. The expected original baseline approval target date may also have been shifted owing to the resource assignment issues described above.

Baseline approval will require trade-off analysis between you and the sponsorship group in an attempt to resolve conflicts in fulfilling the constraints. Project scope may need to be reduced, funding to be increased and/or the schedule duration to be lengthened. Whatever the decision, if the organizational objective and benefits cannot be accomplished, reference must be made to the original business case to ascertain whether the project still has any validity. Often, projects are halted when they are not deemed to bring value to the organization, and the efforts and costs involved in producing the project and product scope(s) are placed in profit and loss—usually the latter.

However, the baseline approval is not the major problem.

5.3.1 Securing Resources for the Project Execution/Implementation

(I omit discussion here on outsourcing, as I deal with that in Chapter 6.)

I trust you have already read Chapter 3 and you have been indoctrinated in the acceptance that projects are performed in the future in a world of uncertainty and unknowns.

In this section I am focussing on *who does what*. I have described the issues and challenges of securing resources during the initiation and planning phases. This challenge is more acute for the plan that covers project execution/implementation, for the simple reason that whatever human resources we have deemed necessary to the project, we have no idea if such individuals exists or whether they will be available when needed. Securing resources for the execution/implementation phase has to be completed before you present the baseline plan to the sponsorship group.

I offer a example of how you might hire new employees. You define a job position and its description and complete that with the skills, competence and experience you seek. You post an advert using whatever media you prefer. And you wait. Some prospective applicants will respond, and your initial job is to vet whether they fit the profile required. From a short list, you then proceed to interviews to arrive at the candidate most suitable for the job. That is not done overnight and may take anything from a few weeks to a few months, and you have to make some concessions when the candidate des not fit the profile exactly. We can always hold out the hope that you may be lucky enough to find a candidate with exactly the profile you want.

In planning for the execution/implementation of the project, you define the complete resource profile(s) needed for each activity, which allows you to estimate its duration and cost. The activity network and the resulting Gantt chart, from a given assumed project execution start date, determines when that activity can start and when it will end on a given calendar. Now you know *when* that resource is required, the *duration* of the assignment and the target *end date* of the activity. You frame the resource need to a specific time period.

Your execution/implementation is based on an infinite, unlimited and unconstrained resource model specifying detailed expertise, where the project defines when an anonymous resource will be required. But where do you advertise?

As discussed above in Section 5.2.2, Securing Internal Resources for the Product Scope Planning, we are faced with the same problem. However, this is more than a problem, as you require a given resource at a prescribed timeframe possessing the exact profile in the future. And you do not know if that resource even exists when you need it.

The problem is compounded when the execution/implementation you are managing stretches over many months or years. How in the world can you even envisage going to a functional manager and requesting, "The project would need this type of resource in month [m] of next year", and not getting a bemused smile as a response (unless the manager in question possesses a superb resource capacity plan and can answer immediately, which is certainly unlikely). Do not think it is easier to secure resources for a project of a duration of a few days or weeks—you will have still the same problem, although most probably less extreme.

As you are in the process of securing the availability of resources prior to getting baseline approval, those functional managers with whom you have to negotiate are part of the stakeholder pool. They should be receptive to the requests, as the outcome of the project dovetails with their own operational needs once the deliverables are produced.

You must go into this resource negotiation stage with a fully detailed project execution/implementation plan, produced by project management software. Following your request for a resource, you can determine the impact to the overall plan if you receive a response which does not match the need. A blatant "No" forces you to either go elsewhere, if possible, or determine with the manager when the resource could be made available. If that works, you then have to update your software plan, as that activity will need to be shifted to later in the schedule, and in a matter of nanoseconds you will be able to determine the overall impact to the project. You will hope that the activity is not on the critical path, as all the succeeding activities will also be shifted, along with the previously established resource loading needs. This is when headache pills will be useful.

Unlike in hiring, where candidates come to you, in the case of projects, you, the hirer, go to the owners of candidates. Upside down world it would seem.

Needless to say, you will need to visit every internal functional manager who may provide the resources you need. And each time you will adapt and modify the execution/implementation plan to fit the finite and constrained resource availability.

Your project execution/implementation plan starts as an *activity-driven* schedule based on infinite resource availability, where you state that these are the activities to perform, now give me the resources when I want them, to a *resource-driven* schedule, where you are told these are the resources available along with their profile, and you have to do with that. Functional managers who eventually inform you that the resources required for the project can be made available must then commit to their agreed assignments; otherwise, you will have difficulty arriving at an honest approval from the sponsorship group, and later you will be on tenterhooks for the duration of the project, as you live in the hope of securing the resources—each possessing the required profile—at the prescribed date. Wishful thinking is the most experienced phenomenon for project failure.

5.4 Actions Required from Management

It is now obvious that projects cannot be launched arbitrarily in the hope that all future work will be performed according to original plans and that all resources will be available for the execution of the activities.

It is pure insanity to think otherwise—at least until management wakes up to the realities of change by projects and that the future is for the most part unknown. Management cannot continue to believe that resources grow on trees or are waiting somewhere to be used as and when needed and that they all have the requisite expertise and availability to perform to a superior level.

When a project fails, tempers run high, upper management will be enraged, project management professionals will be angry, all employees participating in the failed project will be despondent, and the organization will engage meaningless costs. This is insanity.

> Without a realistic and holistic approach addressing internal resource management and the institution of coherent resource capacity planning, projects will continue to fail and fail again and again.

5.5 Chapter 5 Summary

No project can be effectively performed without individuals who have the knowledge, ability and capacity to successfully complete the work that is allocated to them. Work assignment cannot be a lottery, where whoever is available is the appropriate person for the job. Furthermore, any individual assigned to the project may possess some but not all of the requisite skills, or they may even have more experience and skills than required.

Resource assignment faces many challenges for internal work allocation, as it is rare that employees have the required capacity to fulfil the given project task along with either their main job or their assignment on other project(s).

Outsourcing and "body-shopping" on a time-and-materials basis for the project may circumvent the capacity issue; however, the competence and skills element will remain and may contravene confidentiality requirements.

There is no magic wand for project resource management. However, a coherent resource capacity planning system across the organization, and specifically for all projects managed within a project portfolio management system, will provide a realistic and dynamic mechanism to determine the feasibility of meeting any or all of the organization's strategic and tactical objectives.

Chapter 6

The Sixth Sanity Check— Outside Assistance

Rare, if any, is the organization self-sufficient enough to have all the available resources, experience and expertise to manage all the projects of change it needs to launch to achieve its objectives and goals. Most organizations have to focus

Here's what you get when there is a budget restriction on outsourcing!

on what they do best and dedicate their efforts and investments to the areas that matter most. This will invariably set them on the path to a *make-or-buy* decision. Outside assistance is called for in a buy scenario, and the organization, taking on the role of the contractor, will need to pursue a subcontracting avenue to realize its project(s).

There are three main reasons why organizations subcontract part or all their project: they do not know how to do it, they may be able to do it but do not have the capacity at present and/or prefer to invest elsewhere, or they just do not want to do it.

Whatever the reason, subcontracting is a call for help for outside assistance to perform on the project that the organization wants to complete as it seeks to reduce costs, enhance performance efficiencies, access external expertise and gain other benefits. All contractors, vendors, service providers, independent consultants and the like bring their industry expertise, experience and availability of professionals to the organization that requires it, and the contracting organization contributes considerably to those goals.

I am therefore always amazed by those organizations that do not consider the "hands that help them" with more respect and gratitude. Unfortunately, many have an objectionable and archaic attitude towards outside assistance and approach that relationship with a mentality of "I pay, therefore I demand". That to me is insanity.

6.1 The Assistance Providers—The Sellers

Let me use *seller* to encompass all those organizations from whom you will procure products or services, and you are the *buyer*. I have been fortunate to be on both sides, and I would like to share comments and personal conclusions from my experiences as we continue on our journey towards sanity.

I categorize sellers into three types: one-to-many, one-to-some and one-to-one. This categorization matters to understand the principal issues in seeking outside assistance and subcontracting.

One-to-Many Seller

One-to-many is for product mass production: the seller owns the specification, the design and the product and manufactures umpteen duplicates of the product to sell, from rivets to motor vehicles, passing through sweets and candies. The seller has no idea who the final buyer is—in fact, the seller does not really care about the identity; it looks for a market.

One-to-many is extremely specific—you get what is "written on the box", it is off the shelf and has a catalogue price. As a buyer, you should at least know what you want to buy before entering the purchasing cycle. When you decide to procure, you define internally exactly what you want in a specification and describe the type and nature of the product or service—the object. You then go out to find that by either shopping around or putting out a tender to receive bids.

All things being equal, you will most probably choose the lowest price, including financial payment incentives. You may be pressed for time and pay more for the product to be delivered within the timeframe you want. In any case, what you get is an object that you need, and you should have already determined how and where you will fit it and use it. Usually, you would get some form of guarantee if the object is solid or mechanical, and it is up to you to define the boundaries and limits of such a guarantee. Do not try it for a bar of chocolate, unless you can prove the food poisoning came from there.

For one-to-many, the procurement department will often be responsible for the purchasing cycle. They usually have great expertise to achieve the best price when dealing with precisely specified objects.

You will still have to include in your project the activities for lead-time delivery, installation and integration into other components of your project. And consider whatever training and re-skilling may be needed to exploit the acquired object.

One-to-many is more of a purchasing mechanism than subcontracting. Still, you buy what you cannot or do not want to do.

One-to-Some Seller

One-to-some is for integrated systems: the seller owns the specification, design and product and provides and installs the system to be incorporated into the buyer's project scope—think systems like ERP (engineering, procurement and construction) software, aircraft engines, elevators, manufacturing/assembly lines or any system that you need to procure to fit into your requirements.

One-to-some is provided by the seller as a specified integrated system of many components with multiple functionalities. You have established the overarching project scope concept and its requirements, into which such an integrated system would fit its needs, albeit with many integration interfaces. The seller may need to bundle discrete systems from various sellers to be combined into one complete integrated solution to be delivered to you.

You are not procuring an off-the-shelf product, even though that may be how the seller describes it. Beyond having to integrate the purchased system, the rest of the project scope still has to be fulfilled. This will be done in conjunction with other concurrent projects and whatever organizational readiness preparation and transition is required for the acquired system to be functional.

In large programs you may well have multiple system integrators as your sellers, and the effort of coordination between these and the receiving functional departments will fall upon you.

Another issue will be to define the functional guarantees that the different system integrators can and will provide, as well as setting how and when the support clauses are to be triggered.

You will also need to decide with the system integrator seller(s) whether contractually you become the outright owner of the procured integrated system or a licensing agreement is to be established.

One-to-One Seller

One-to-one is for the delivery of a custom-built solution—the buyer owns the specification, and the seller is called upon to produce a unique solution for the buying organization. One-to-one is in the realm of turnkey projects, where the solution is to be built, supplied and installed complete and ready to operate. Regarding solution delivery and approval, there is transfer of ownership to the buyer, and the same solution must not be found elsewhere, unless an agreement is drawn with the seller.

One-to-one is driven by what the buyer wants exactly, albeit without specifying all the fine details. It will be unique, as it corresponds only to the buyer's wishes and desires. The buyer will need to define the needs and requirements and initially request a design from the seller for a corresponding solution.

When the seller is to provide a design prior to actually building or fabricating the solution, this then engages the buyer and the seller in discussions and negotiations as to which organization will fund the front-end engineering and design. In the construction industry both organizations may opt to proceed with an EPC.

The one-to-one solution's defined scope will be contained in a broader scope for the organization. For example, the seller's delivery of a totally revamped and new technological infrastructure or a new corporate headquarter is only a part of a wider scope covering and affecting the organization, and there are many more examples possible. In all these cases, you will still be managing a project, or most probably a program, which must ensure that the turnkey solution provided is exploitable.

There is therefore a wider and more extensive project scope content to be delivered to the organization and its functional departments. For example, the transition to operations from one IT infrastructure to a new technology structure and platform will certainly affect all digitalized processes, and their migration and will require user training for all staff affected, to name the major impacted areas. For a new corporate headquarter, the project must include the plans and execute the relevant activities to ensure that the organization maintains its operational function as staff are relocated and reskilled with the corresponding training.

Such a move must also consider all other infrastructural changes and transitions required. All these extended scope requirements needed by the organization may well not be included in the one-to-one solution brief and contract, and thus will be considered out of scope for the seller.

Following delivery of the turnkey solution, the buyer organization becomes the owner of the delivered results and can operationally exploit them. The seller may have contractually agreed on a ramp-up operational period to provide support and assistance to the solution. This would not usually include post-delivery support to other internal functional departments within the global scope of the project and deemed out of scope by the one-to-one seller.

> The organization has to differentiate between the types of sellers and how it engages with them.

6.2 Do Your Homework!

From this point on in this section I will only concentrate on one-to-some and one-to-one project solutions.

Your organization needs a solution to achieve its goals and objectives, and you have been elected as the project manager who will accomplish that. Now you have to do the homework for the project. You would expect that a full business case analysis has been done and that the organizational reasons for this project are well understood by all major signatories. An initial make-or-buy analysis will determine what you may need from outside assistance.

On completion of a comprehensive analysis of the project's master needs and requirements, including a high-level design, you will determine what to have done outside, and you enter the buy scenario for the portion of the requirements to be subcontracted. You now have to define the boundaries of what is to be subcontracted, and you will need to specify its contents with as much detail as the current scope of the project allows. The remaining scope requirements will form part of the core project contents that will be performed internally under your responsibility.

You may be in a situation in which you require more than one one-to-some sellers, and you then have to define how their interfaces will be specified, as well as the links into your core project scope.

When opting for a one-to-one seller, you will also have to define how their scope of work interfaces with your project. I do not recommend that you deal with more than one one-to-one seller, unless you have to or you like challenges.

Depending on the nature of your project and the needs and requirements you have specified, the detailed non-subcontracted design for the project remains. Remember that you are responsible for the totality of the project, even if you will subcontract a large chunk of it. You still have to design how to integrate the subcontracted requirements and the corresponding solution into the operational departments and functional entities, infrastructure and processes of your organization and achieve the results of your project mandate.

Functional requirements and design for operational transition, migration, training, exploitation start-up and operations continuity will rarely be within the scope of the work you subcontract. You also have to consider other on-going projects and how your project fits in with the master picture. Think Legoland, but you still have not produced all the pieces.

The subcontractor will subsequently be provided with the functional requirements as an SOW (Scope of Work) and should also be provided with the project's high-level design. On receipt and review of their proposed solution, you will have to map their design into your overall design, which is a major challenge. More about that below.

> Your homework consists of establishing
> the master project requirements and defining
> the statement of work for the seller.

6.3 The Search for the Suitable Subcontractor

The identification and selection of the subcontractor is a procurement process which your organization should already be using. Most probably that process is well honed and efficient for one-to-many purchases and your organization has already established a preferred vendor list.

However, the process may not be geared for one-to-some and one-to-one procurement.

As with all types of procurement, we enter the three-letter acronym world of the RFI, RFQ, RFT, RFP: request for information, request for quotes, request for tender, and request for proposal, respectively, as well as any other acronyms you may be using.

Subcontracting proceeds along these steps to finally arrive at an acceptable technical and financial agreement with the external organization.

6.3.1 Establishing the Statement of Work—The SOW

The journey to arrive at the subcontractor selection begins with the comprehensive formulation of the SOW: the expectations and the deliverable(s) to be achieved. You must state the constraints you impose on scope contents and, where necessary, schedule and budget.

Furthermore, you must clearly specify the selection criteria that will be used to select the subcontractor and the documents they must provide for you to assess the technical and financial contents. You must also define the response time given to the subcontractor to provide their proposed solution.

The subcontractor must provide a detailed, designed technical solution and financial proposition, along with a project execution/implementation delivery schedule based on an assumed start date, including the resources to be made available by your organization and the interfaces to operational departments and concomitant projects. It is imperative that the subcontractor specify and itemize the risks that will be covered, the out-of-scope contents they will not cover and the assumptions on which their proposal is based.

Procuring a one-to-some or a one-to-one solution is definitely not an off-the-shelf purchase, and the challenge is determining which subcontractors are not only capable of responding, but also their availability within the timeframe you require.

Your organization may already have a preferred vendors list based on past contracts and satisfactory performance, or else you will need to conduct due diligence and a viability analysis of the subcontractor prospects, which may not result in identifying the most adequate organization for the SOW.

Depending on the scope contents to cover in your SOW request, you must dedicate a timeframe to conduct the RFI and establish a first short list of prospects. Once selected, you will request the short-listed subcontractors to provide a tender or a proposal. For public organizations, you will also have to conduct public meetings with all prospects to answer questions and to clarify certain areas of the SOW. You may also conduct such meetings if your organization is in the private sector.

All these steps will be time consuming and require you and potentially certain core team members to be assigned to these tasks. And you still, on one hand, have not received any offers pertaining to the SOW, and on the other hand have not assessed and analyzed the proposal contents.

6.3.2 Assessing the Technical Proposal

You must concentrate principally on the technical and designed proposed solution. For intricate and complicated SOW scope contents, and certainly for an eventual

EPC contract, you have to establish and determine whether the subcontractor(s) will develop a designed solution at no cost.

You cannot proceed with any contractual agreements with a selected subcontractor unless you perform a detailed analysis of their technical proposal. Your analysis must extend beyond what is proposed by the subcontractor(s), and you need to determine how their solution fits into your overall project and the different functional departments that will be the recipients of that solution.

There may be areas where the subcontractor's solution design does not fit the overall project design, and changes to their proposal would need to be incorporated. This may set off a series of interchanges that may or may not be acceptable to certain subcontractors, causing them to no longer wish to pursue the proposal phase. For those subcontractors who may wish to engage in adapting their design to align with the changes, the question of how to cover the additional costs to them will need to be discussed and agreed.

6.3.3 Assessing the Financial Proposal

You should not consider the financial proposal until you are satisfied that the technical solution proposed is viable and fits into the overall project. Now opens another type of assessment and analysis.

When dealing with a one-to-some solution proposal, the core financial portion may well revolve around the cost of the physical object(s) or for application software for the cost of licenses. In addition, the financial proposal must be assessed to determine all the other activities related to the integration and subsequent exploitation of the solution.

- For physical objects, such as mentioned above, engines or assembly lines, activities for delivery, installation, insurance, system testing and, when necessary, regulatory certifications need to be included in the proposal. Training and solution support costs must also be considered.

- For software application solutions, such as ERP, CRM or others, these activities may cover the transition from old to new systems and data, training for users, operational changes to current processes and the post-deployment support on offer. Furthermore, you must ensure that the interfaces with other one-to-some solution proposals are well defined and coherent within your overall project scope.

When dealing with a one-to-one solution proposal, you are to procure a turnkey project solution. Rare will it be that the different solutions proposed by the different subcontractors can be assessed only on a financial basis. The fundamental nature of a one-to-one solution is that it will be specific to what the subcontractor proposes, and the designed solution will certainly be different to its competitors.

How can you evaluate the cost of two completely different designs that have been based on the same SOW? Examples abound in the construction industry for commercial building solutions.

6.3.4 Assessing the Schedule Proposal

Remember, all schedules are provisional and speculative views of the future, and thus are conditional on events outside of your control. This is aggravated when the timeframe of your overall project and that portion attributed to the subcontractor stretches over many months or years. Thus, before any analysis of the proposed schedule, you must review the risk events covered by the subcontractor and the assumptions that have been identified. More on this below when establishing a contract with a subcontractor.

The schedule proposed by the subcontractor will become a subset of your project. Whatever your choice, you must assess the proposed timeframe and schedule, which would be based on an assumed start date as stated by the subcontractor. The schedule is to include the resources that your organization must provide for the implementation, roll-out, transition and exploitation of the solution.

You will need to synchronize the subcontractor's schedule with your project's core activities as well as those from concomitant projects. This analysis will extend to those of the functional departments' schedules, which may demonstrate lack of availability of their own resources or an incompatible timing that coincides with their operational peaks and/or end-of-year operational closure periods. This may well cause changes in the milestone must-dates for the subcontractor, and these will need to be resolved. The consequences for the overall project schedule must be assessed.

> Do not choose a one-to-some or one-to-one proposal on price only, as seller's solutions will differ.

6.4 What Is in Your Contract?

Let's start with two definitions regarding "contract":

"A contract is an agreement between parties, establishing mutual obligations that are enforceable by law. To be legally enforceable, a contract must include certain elements: expressed by a valid offer, formal acceptance and mutual assent, authority, adequate consideration; capacity; and legality".

"A breach of contract is a failure, without legal excuse, to perform any promise that forms all or part of the contract".

There are many excellent books, articles and papers on project contracting. I will not venture into the intricacies of establishing and agreeing contracts with sellers. There are innumerable lawyers and financial experts who possess great expertise in this domain and practice their skills daily to ensure that every clause in a contract protects its client. I encourage you therefore to read the large amount of information already available elsewhere.

I will also not enter into a discussion that describes the actual contents of any contract with a seller, in this case the subcontractor for a one-to-some or one-to-one solution. Nor will I explore who actually drafts and will negotiate the contents of the contract and its type. The only comment I will make is that the enterprise procurement department, perfect for one-to-many purchases, may not be adapted to this task. Furthermore, the contracting rules may be very different whether you are part of a private or a public enterprise.

Rather I want us to delve into what the actual contract is about: a proposed solution for a given statement of work. I wish to consider with you what goes on behind the scenes and address one-to-some and one-to-one contracting, as the contents of the required scope are not precise and detailed, whilst usually the one-to-many purchase contracts are exact. There are many situations where I will highlight what I consider to be part of my treatise on insanity.

When you subcontract, you will have established a contract: fixed price, target cost with or without incentives, and time and materials. Subcontracting in many cases means you do not know how and what to do. You rely on the expertise of the subcontractor. How do you verify that?

Bringing in project management services doesn't mean handing over the reins of your project to the subcontractor; rather, it means making sure your overall project comes to life as you are intending. Bringing in external experts also means you can stay in control of the things you know best and not have to second guess what you don't.

Subcontracting companies look for profit. They bring expertise and resources and ensure compliances are followed and certifications obtained, along with managing legislative constraints.

6.4.1 Fixed-Price (Lump-Sum) Contracts

For a one-to-some and one-to-one solution, a large proportion of such subcontracted projects will be based on a fixed-price (lump-sum) contract. Invariably, public organizations will be using such a contract. In some cases, you may opt for flexible fixed-price contracts, with a specified target price and ceiling price, giving you a firm price limit.

However, what does a fixed-price contract actually mean, and what do you actually fix? You have to establish the contract to revolve around six major vectors, as follows:

- Obviously, the first thing that comes to mind is to fix the total price you will have to pay. But that is not all that has to be fixed.
- You have to fix the subcontractor's scope contents, and that is the combination of the statement of work you initially provided and the proposed solution design. The fixed scope contents must integrate into your overall project scope.
- You have to fix the schedule. The proposed subcontractor schedule will assume a start date for their work and indicate the intermediate milestones and the target end date of their intervention. The schedule proposed must be synchronized with your project's master schedule as well as the different schedules established by the receiving operational departments.
- You have to fix the subcontractor's coverage of risk events. These must be described in the contract, describing in detail the risk events and the responses to them that will be managed by the subcontractor, so as to ensure that the risk transfer to the subcontractor is coherent.
- You have to fix all the clauses with respect to the terms and conditions of the contract. This part is obviously the most important in the contract and best left to legal professionals.
- And finally, you have to fix the management relationship between your organization and that of the subcontractor. Beyond identifying the subcontractor's project manager, the relationship management will also involve decision-making milestones and the participation of equal-ranked individuals from both your organization and that of the subcontractor.

6.4.2 The False Security of Fixed-Price Contracts

There is a temptation and insanity to consider fixed-price contracts as secure and easy to establish, with the added value that "someone else is taking care of business". These contracts are fraught with dark sides. Hovering above these dark sides is the constant reality that you cannot permanently fix any of the six vectors described above, as all are subject to change in an environment of change, just like long-term weather forecasts.

- The fixed price makes one side a winner and the other a loser, and the two are irreconcilable. You seek a fixed scope delivery, within a pre-determined schedule and a fixed price, whilst the subcontractor principally seeks profitability. However, the subcontractor, who is expected to fund the project costs, will absorb and take the burden of their own increased costs when the statement of work has been misunderstood, technical issues were not correctly addressed and there are design failures, resource needs and risk events were not properly assessed, plans and schedules were created optimistically, and cost estimations have been incorrectly calculated, to state the major reasons.

The consequences to the subcontractor are to slip into a scenario of rising internal costs and negative profitability, uncontrolled schedule slippages and scope delivery under-performance. The major consequences to your master project will be overall schedule delays and failure to fill the core project functional requirements. Expect lengthy legal disputes and eventually the contract abandonment of your chosen subcontractor. Which heralds yet another lengthy delay, with an increase in your associated costs, since you are now forced to seek out a replacement subcontractor.

- In a fixed-price contract, the subcontractor has presumably calculated their own project costs and profit margin. However, budget inaccuracies will always exist and are made worse when insufficient hedging on raw materials and other resources has not been included in the price proposal to consider market pricing evolutions above a given inflationary rate.

  As costs rise and profit margins dwindle away, the subcontractor may no longer be able to complete the work within the original price and may even forfeit the contract when the penalties are judged to be acceptable. This obviously leaves you in a bind.

- There will always be changes in a project's scope, a factor that has to be acknowledged by both parties. This is made more evident when project durations are expected to stretch over more than one year. Changes will come from both parties, as well as from external legislative bodies. Your master project scope will need to address internal changes required from the organization, upper management, user groups and concomitant projects.

  Hopefully you have instituted a change request management process (discussed further below) to assess and measure the impacts to your project's objectives. Adjustments will have to be made to the original statement of work agreed with the subcontractor. This is further compounded when changes provoked from internal sources occur in the latter stages of the delivery schedule, and previously delivered portions of the project have to be either reworked or even abandoned. Negotiations with the subcontractor to entertain changes to their statement of work will undoubtedly lead to contract clause negotiations, which may often make the original fixed-price contract no longer valid.

- Fixed-price contracts are usually more challenging to the subcontractor's project management team. Constant supervision and evaluation of the actual scope performance and cost of their project have to be robust to alert them when they will no longer have any profitability. Lack of project management expertise will inevitably lead the subcontractor to make tough decisions for the survival of their contractual obligations.

A lack of vetting of the subcontractor's project manager and supporting project team prior to concluding the contract, and your adopting a rather distant approach to overseeing their performance, will often lead to both parties' working in the dark as to the other's progress. Lack of coordination between the parties' project managers will cause enhanced frictions and sour interchanges when facing problems and issues.

- Relationship management between you and the subcontractor is of utmost importance. I discuss this below in Section 6.6, The Human Side of Sub-Contracting. Suffice it to say for now that from the outset of an agreed fixed-price contract, both parties are, maybe naturally, concerned that each can fulfil the contract obligations.

A subcontracting enterprise is not perfect and may not be equipped and resourced to fully comply with its contractual commitments. Its internal functioning and management decisions may well lead it to accept a contract knowledgeable that their profit margins may be low or even non-existent. The consequences are obvious—under-performance and higher costs from the subcontractor, lengthy negotiations with your organization to determine what needs to be modified in the original contract, various impacts to your master project's scope, schedule and budget, to name just a few.

The major problems with fixed-price contracts are the result of a conflict of interests between you and the contractor and the fixed scope of work. Contract price adjustments will certainly test and damage the relationship, resulting in a lower quality of the delivered outcomes, which will only generate increased costs to your organization following operational release. Eventually, the relationship between the parties will change for the worse, which is of no help to anyone.

> Fixed-Price contracts are NOT a panacea.

6.4.3 The Contract Is Not the Solution

In a large number of the projects you subcontract, much effort is to be placed on the design. How much knowledge and expertise do you and your team members have to understand and verify the technicality, soundness and applicability of such a proposed design solution to your project's needs? How do you assess that the deliverables defined in the subcontractor's solution can be physically produced, tested, verified and accepted? What safeguards are in place in the contract?

When you agree on an EPC contract, the subcontractor is responsible for all activities in the Statement of Work, from design, procurement, construction, to commissioning and handover of the deliverables to the owner or operator. EPC contracting will be a subset of your complete project scope.

I have rarely found Specifications of Work—or any other specification document agreed with a subcontractor—to be perfect. Remember, there are Knowns and Unknowns and the assumptions that accompany them. There will be misinterpretations, omissions and maybe big holes in a specification brief. The worst is when these are not highlighted during contract negotiations (maybe they should be by the subcontractor), and what is the consequence? Change requests, conflicts, litigation and all the wonderful situations that make lawyers rich. You must therefore have a robust change request/order process instituted, shared with your subcontractor and described in the contract.

Adjoined to the contract you must be in possession of a clear and detailed view not only of the scope contents to be delivered, but most importantly of the subcontractor's execution/implementation schedule and how it dovetails to your master schedule and the resources loading required by your organization. Furthermore, you must be knowledgeable of the subcontractor's selection of their contractors and suppliers to their proposed solution and how their corresponding delivery schedules align to the overall project schedule.

Too often, delays in delivery are caused as a result of multiple suppliers under the responsibility of the subcontractor's scope or the insufficiencies of your own organization's resource availability. And this falls back on you and gives excellent reasons for the subcontractor to justify their own delays.

When you contract a fixed-price or lump-sum contract, this means you are transferring the associated risk events contained within the scope of the statement of work to the subcontractor. It is incumbent upon you to review in detail the risk events that the subcontractor will cover and the assumptions that have been identified. *Force majeure* or government legislative change events and how they are to be handled must also be clearly defined in the contract. Are they?

Final delivery and acceptance/commissioning testing of the subcontractor's solution are mired with functional defects, quality problems, schedule delays, resource availability issues, commissioning and certification setbacks, all creating an even more tense and stressful environment as your organization prepares to go operational.

> The proposed subcontractor's technical and financial solution is what your procure. The contract between both parties only safeguards their interests.

6.5 Are You Being Delivered What You Expected?

The litmus test of your subcontracted project is done during the execution/implementation phase. You cannot be expected to be in multiple places at the same time, especially if your project requires many subcontractors, suppliers, service providers and the like, on top of the internal organization communications and interfaces you must conduct.

So—how do you know if what is delivered from the subcontractor is what you expected?

You will of course conduct the institutionalized weekly and monthly classic project meetings and quality reviews, which often may not give you the true picture of things, especially when the average duration of activities does not synchronize with that given frequency, as some tasks may be of one-day durations whilst others may need to be completed over many weeks. For the former, you are dealing with historical data which may have needed immediate responses, whilst for the latter you cannot truly determine the actual progress. Beware the use of earned value management for schedule progress analysis—you must conduct upfront detailed planning to derive any senseful information. The schedule variance is often imprecise and at times meaningless and is of potential use only for tasks that are performed at a linear productivity rate, as progress is measured on percentage of actual work done.

- Projects where physical objects are created by the subcontractor during the execution and implementation are easy to see, and their progress and status can be assessed on a particular day, as you cannot cheat the calendar. However, their functional quality requirements cannot, until testing and approval is conducted. As for the contractor's associated costs incurred to date for a fixed-price contract, you can only hope that they still have some margin of profit.
- For non-physical technology-driven objects, their progress is far more difficult to assess until the results emerge from their depths. Progress for such subcontracted projects unfortunately will rely on the famous 90% Complete Syndrome—which often does not mean there is still 10% of work still to complete. For these types of projects, functional quality requirements cannot as yet be fully tested until a substantial part of the solution has been created, and the same comment can be made for the subcontractor's accrued costs to date and whether they will still remain in business.

Human nature is by definition in a constant protective mode, especially when things do not go the way we want, and we have to honestly report on our status. There will always be a feeling that certain reprimands may be forthcoming when we are not in line with scope contents, schedule and costs. We then have to decide whether to enter the Volcano Syndrome mode. We can infer when this mode is entered when a phrase like "We are encountering minor difficulties, and we are working on that". Sounds familiar? The Volcano Syndrome is when an individual may sit on a problem, hoping it will be resolved at a later stage. It is like asking someone to sit on a volcano hoping to snuff it out. However, nature is much more powerful than your backside. And the aftermath may well be catastrophic.

6.5.1 Projects Are Dynamic—Expectations Will Not be Met!

The biggest pitfall in subcontracting is not only the seller's inadequate performance. It is the somewhat distant attitude of many parts of the organization, thinking, "let the supplier get on with it, and if they are late not only will we penalize them but also have an excuse for our own lateness". That attitude cannot give you a solution that is needed to accomplish your overall project objectives.

There are a multitude of problems when subcontracting, which often places organizations in a cautious and suspicious mental framework. Problems such as conflicts and disagreements between contracting parties, lack of compliance with the scope of work, delays in project activities, and milestone payment disputes are just a few that abound in subcontracted projects. And it is reasonable to understand why organizations are wary about the approach.

Experience has shown that no subcontractor will deliver exactly what was contracted. The principal reason is that, as I have mentioned many times before, a project will be performed in the future in a volatile and changing environment.

Market evolutions associated with internal organizational changes will lead you to review the overall objectives of your project and have repercussions on the subcontractor's cope of work. Change requests from both parties will happen, and the structure of the common contract will be strained, at times to the breaking point.

When subcontracting a solution which is novel, be it in its architectural or technological design, you must expect that scope contents will evolve and change.

As a football fan, I always have in mind the delays in the construction of a new iconic stadium in London, from the delays in the demolition of the old stadium to the novel and untested construction of the arch spanning the whole length of the stadium. A litany of other problems was also encountered, such as major design changes, poor performance, poor site management, insufficient risk allocation, worker protests and strikes, etc.

Adversarial contracts and litigations added to this pile to provoke cash-flow problems. The delivery schedule was initially earmarked to start before Christmas 2000 with the demolition of the old stadium and to be completed during 2003. As a result of a variety of delays, the demolition began in September 2002, and the new stadium construction started in 2003, targeted to be completed in 2006. Eventually the new stadium was officially opened in March 2007, more than three and a half years late, or 100% schedule delay. The initial budget was £362.5m and ended at £757m, for an increase of over 200%.

Which unfortunately proves once again that pharaonic projects will never deliver to the original contractual conditions. However, brought to a smaller scale, your subcontracted projects will also suffer in the same way and blow your contract to smithereens.

6.5.2 Be Ready to Master the Change with the Subcontractor

You have the sole responsibility for the master project and the organizational changes to be instituted by the project's results. In a one-to-some or one-to-one solution, the subcontractor plays a large part in your project's success. You therefore must be on top of the overall progress and monitor all deviations, variances, and scope changes that occur and take the appropriate decisions to achieve your project's goals.

The biggest problem you face with a fixed-price contract is that you are working with a Black-Box Syndrome, where the focus is on what you are obliged to pay on subcontractor progress at established milestones. Few are the occasions when the subcontractor will share details of schedule and costs in such a contract. However, the most difficult part of the fixed contract is the statement of work content that is performed and delivered by the subcontractor and how you verify that.

As will be normal following a classic project management monitoring and control approach, you must principally concentrate on the progress and evolution of the master project's scope contents, its schedule and the associated accrued costs.

Scope contents will certainly evolve, and I have explained sufficient times as to the reasons why. The evolution will originate either from your master project or from the one-to-some or one-to-one solution provider.

- Scope changes initiated by your organization will have repercussions on the master project. When these changes affect the statement of work you have subcontracted, you must determine how this will affect the subcontractor's commitments and the related contractual price and clauses. Similarly, the contract has to be reviewed when the subcontractor forces changes to the statement of work scope contents. Evolutions which are forced upon the project from external sources, such as legislative changes, and how they are to be addressed must be described in the contract to avoid litigations over who is to pay for the additional work.

- Change requests and orders are extremely important to manage. You must have a well-defined, robust and mutually accepted change request management process instituted and employed by all parties, and this must have been referenced in the contract. At issue is determining the origin of the change request, be it your organization, that of the subcontractor, or from external government bodies. Special clauses must be included in the contract to cover natural disasters and *force majeure*. The contract must clearly define in what cases which party is to pay for the change(s) and how any delay imposed on the project is to be managed. Contract updates must then ensue, which may cause further delay.

Project schedule delivery delays are the most visible, as you cannot cheat the calendar when milestones are not met. There are many reasons why scheduled dates are not met, and I have explained this in many parts of this treatise. There are multiple repercussions to your master schedule when delays originate from either party. You will need to assess and review the current status of your project and what changes have to be made to accommodate the imposed delays and the re-distribution of resources for the remainder of the project. Your organization's functional departments will also need to re-schedule their own plans for the operational start-up of the solution.

When delays originate from the subcontractor, you will need to review and redraft the contract. This may lead to early termination if the subcontractor is no longer able to complete the work.

Your master schedule will undergo many internal changes due to requests to modify the project's goals and objectives from key stakeholders and operational managers. Changes requested in the latter stages of your project have the greatest impact to the contractual clauses, as these may cause a complete upheaval of the subcontractor's ability to continue and may even force the contract to be nullified unless the requested changes are rejected. Three such occasions come to mind:

- The CEO on a visit to assess the progress of the new HQ building decided that it would be great to accommodate another business unit in the building, requiring two additional floors.
- The SVP of Personal Insurance for a major broker was invited during systems testing to a demonstration of a new insurance policy application, then wanted to change more than 30% of the screen and paper reports, as he "had an idea."
- The recently hired and on-boarded head chef at the near completion of a hotel for a major chain imposed a totally new design of the kitchen.

In all cases when projects have to be re-scheduled, you must expect contract negotiations to be conducted with the subcontractor and an increase in both the costs and schedule of their and your project.

Scope contents deliverables from the subcontractor require acceptance after verification and validation. This may involve a variety of internal and external entities to conduct the quality control necessary. At issue will be the potential delays caused by your organization and that of the subcontractor during the approval cycle timeframe, if and when the master schedule has not included the elapsed time necessary for the deliverables' verification and validation cycle. Delays may be further compounded when legislative, health and safety and other formal certifications are to be given by local government bodies.

Formal acceptance of subcontractor deliverables will also need to consider items in the various defect lists, punch lists or snag lists, which all need to be resolved either immediately or post solution delivery. Thus, delays in the approval of subcontractor deliverables may create a spill-on effect on the overall master schedule.

> Remember, one-to-some and one-to-one solutions never deliver to the original scope, budget, schedule and contract.

6.6 The Human Side of Subcontracting

A project brings together people from different organizational entities possessing different expertise, competence and experience. You are managing a social group and a community of individuals from diverse backgrounds with personal objectives.

You may be on the same wavelength with members of your own organization, as one may assume that eventually all of you are supportive of your company and wish success across the board. In subcontracting, you also include in the social group people from the seller organization, and they would most probably have the same sentiment towards their own organization as the people in yours would have.

The biggest insanity is to have "we against them" mentality.

Should there be a clash, with both parties on the opposite sides of an imaginary fence, or even a wall? Subcontracting will certainly increase that. When you, as a buyer, entertain a somewhat superior stance towards a subcontractor, you will unconsciously place them in a seemingly inferior position. And you may be manifesting that behavior just because you believe that you are the one to dictate what you want, and moreover, you are paying for it.

Do not forget—the subcontractor is part of your team. They are not outsiders. Bring them in, create a common integrated team with the same goals. Of course, subcontractors will protect their own organizations—that is normal. You will protect yours as well. When both parties understand that, they can mutually accomplish success together.

Both your organization and that of the subcontractor have correctly and rightly established a contract of work. Fine.

Contract contents are never complete to the smallest detail and totally comprehensible to all readers for the many reasons I have cited in previous sections. Furthermore, the somewhat static form of a contract will always be challenged and bent owing to evolutions in a dynamic future that we do not control. Changes will happen, and rather than spending time and money in conflicts and litigations, thanks to lawyers with financial appetites, it would be better if, by a mutual respect of the opposite party, holes can be easily filled. Unless of course there are blatant changes as a result of unprofessional understanding.

In a fixed-price contract, clause-related problems arise. You need to engage in honest communication with the subcontractor and discuss your mutual objectives. A dependable seller will help you reach those, even if that translates to them a project with higher risks and a potentially lower profit.

Otherwise, when both parties are cautious and suspicious of each other, even a comma in the wrong place in a sentence can lead to arbitration. I invite you to read about the missing Oxford comma.[*]

> One-to-some and one-to-one solution providers
> are PEOPLE. Treat them as such.

6.7 Chapter 6 Summary

When seeking outside assistance, the project organization will be dependent on the success of an external entity. The project manager now becomes a principal stakeholder for the seller organization, and the statement of work agreed between both parties contains the needs and requirements for the outsourcer project scope.

Outsourcing entities must be assessed prior to closing an agreement and contract. The project manager must collaborate with the central procurement to ascertain the contents of the technical proposal, not just the price. When not dealing with off-the-shelf product procurement, an assessment of the management structure and financial capacity and status of the outsourcer is of prime importance.

The project manager must assess the outsourcer's understanding of the need as described in the statement of work, the technical approach to be pursued and the technical capability to be made available to the project. This includes a

---

[*] https://www.abc.net.au/news/2017-03-21/the-case-of-the-$13-million-comma/8372956

description of the various human resource profiles that are to be made available to the project.

The project manager, along with the procurement department, must vet the outsourcer's project management approach, past performance in the domain with references, along with their ongoing performance capacity and availability.

When engaging in one-to-one solutions, special attention needs to be placed on intellectual property rights and proprietary rights.

A robust change request/order management process must be in place, as all projects will undergo dynamic evolutions, which are reflected in the master schedule. Avoiding litigation is a constant concern for the project manager.

Aside from the somewhat dry aspects of contracts and statements of work, with accompanying plans and progress reports, the underlying factor is that *people* engage in a collaboration to achieve results for the project, and relationship management is of utmost importance.

Remember: *You* are seeking help, and you are *all* in the same boat.

Chapter 7

The Seventh Sanity Check— Engaging the Efforts of Others

There are myriad books and articles on management and leadership. Observing what goes on in both private and public organizations globally, I can only conclude that there seems to be a worldwide pandemic of illiteracy. What is proposed in these books, to which I fully adhere, has a strong Western bias and is often not practiced in everyday work environments, be it in the West, East, North or South.

Without going into a discussion on belief systems, humans, as defined by *homo*-something, have existed for millions of years. Humans are communitarian, and as such perform better in groups. All groups function according to a hierarchy, be it imposed by an individual or chosen by its members. According to anthropologists, without a structured hierarchy in which individuals focus on the collective good of the group, it may be that humans will have become extinct.

For several years I had the opportunity to consult with and deliver training to an organization from the energy industry. I travelled to all continents except Antarctica (yet) and spent many weeks per year delivering my services. As part of a small group, we were also tasked with presenting management of projects in various international locations to executives at VP-and-above levels during a scheduled morning session. This organization engaged many billion dollars per year in CAPEX, so one could imagine that they should know what to do in their field.

I always enjoy communicating with VPs, as they seem to be viewing the world from so high above the rest of us. I compare a VP to an admiral of a fleet of vessels, overseeing the ballet of different ships bouncing around in the ocean. Admirals have a strategic and total view and have the responsibility to guide the fleet to its goal. Do they have any idea of the sailors toiling in the engine rooms?

My presentation slot was to be one hour, and I needed to be concise, brief and to the point. I needed to determine how I could engage with a group of VPs who were responsible for such large sums of money that financed their programs and projects. So, I went into my usual provocative style.

After the usual welcome and exchanges with the participants, I started with a simple question: *"What is the first thing you think about when you arrive at the workplace?"*—a simple enough question to my mind. In return I received answers covering all types of financial measures, operational performance measurements, profitability, share price, status of strategic development. It was not really the answer I was looking for as the *first* thing to think about.

So, I asked again. This time I received different replies such as environmental impacts, health and safety issues, the tally of work-related accidents, impact of changed legislation, current customer's experience, trends in sales, innovations in the pipeline. This went on for a few more rounds, then I gave up and asked another question: *"And how do all those important things actually get achieved?"* Finally, amongst another set of answers, I got to the word I was looking for: *people!* Yes—all the above cannot be realized without people—the lifeblood of any organization.

I must admit, I got some quizzical looks, as if all their answers had not been valid. On the contrary, they were all valid; however, people have to be the first preoccupation of any person responsible for a group of individuals in the organization. So, I summarized, so as not to rile the audience even more, that from a supervisory level to the highest echelons of management:

> All your objectives, goals and results cannot be achieved without the efforts of others—**people**.

What a simple statement. This sanity is the most important to promote, as the workplace houses a community of people who all have extraordinary abilities as human beings, and they all seek to satisfy a variety of needs (thank you, Mr Maslow).

7.1 The Workplace Community

Let me further delve into the workplace community and invite you to play the "Workweek Monopoly Game" (WGM). I believe we all know of the board game Monopoly and that each time a player passes Go, money is collected. In WGM, every time a person passes midnight, they collect 24 tokens of one hour each. During the five-day workweek, each person will "spend" their tokens.

Each day's tokens will be spent on basic physiological needs, such as eating and sleeping—for argument's sake let us assume that will be 8 hours = 8 tokens. Next, there is bodily hygiene, such as washing and maintaining bodily functions—let us assume that may be 1 hour = 1 token.

By now nine tokens have been spent on the obligatory essentials to be able to pass the midnight Go box again.

We are left with 15 tokens.

The vast majority of us work for our income, and we have obligations, usually outlined in a work contract. We are therefore obliged to travel to and from our home and the workplace (those of us who work remotely may shuffle to our home office area). The time needed for travel will certainly differ between those who live in the cities and those who live in the suburbs or even the countryside. Let us assume that a large majority live in or around large metropolitan areas and take 2 hours = 2 tokens for the back-and-forth journeys.

We have now spent 11 hours = 11 tokens, all on obligations. That is close to 46 percent of our day gone. These cannot be in any way hours spent in the "pursuit of happiness".

From the 13 tokens remaining, we will spend tokens in the workplace. As a standard workday consists of eight hours, plus a one-hour lunch break, we will spend 9 hours = 9 tokens.

> Now here is the crunch: close to 70% ($^9/_{13}$) of our daily workweek is spent in a building that is not ours with people we might not otherwise have chosen to be with!

For the people we love and choose to be with and the place where we live, the vast majority of our time is spent on sleeping and satisfying our other physiological obligations. Wow—what a killer. We only have four tokens left for the rest of the day.

The obvious conclusion is that it is in the workplace that the social group, composed of many heterogeneous individuals, must collaborate and function in unison to have any chance of achieving their common goals and objectives. And that is the prime focus and responsibility of any manager.

7.2 The Responsibility for a Social Group

Now, this brings us to the main protagonist of our story—the project manager.

As a project manager, you are placed in a supervisory role, and you will have been given the responsibility to accomplish results that the organization needs to exploit. You are now part of the hierarchy, and team members will report to you, albeit temporarily. You will report to a sponsorship group and relate with other managers, employees and external providers, who are all project stakeholders. This ephemeral social group constituted by the project obliges you to demonstrate and apply human qualities of management and leadership, as you must engage the efforts of others and their participative commitment to achieve the goals that have been set.

Sanity requires that the organization provide all project managers with the competence and skills to orchestrate, coordinate, facilitate and animate a social group of diverse individuals who may all have different life objectives but happen to be in the same place at the same time.

Further to being trained and adept in the hard skills of project management, it is more important that project managers be competent in the soft skills of leadership and managing people.

A simple rule is:

> To be a leader you need followers, and
> it is the followers who decide.

**I suggested he should
read on Leadership**

Project Leader

Therefore, it is what you do and how you act that will generate the other person's willingness to follow.

Setting direction to reach goals and inspiring individuals to pursue these goals are the fundamentals of leadership. You are faced with articulating the project's path into the future, and you need to be eloquent and impart meaning to others who may not have a clear picture as to where they are going and why. You can take inspiration from Jonathan Swift, who wrote, *"Vision is the art of seeing what is invisible to others"*, and make tangible and significant the journey to be followed.

A project provides you with the ability to set the direction, as that is clear in the mandate and objectives you have been given. It is then up to you how to formulate and present this to all project participants. This calls upon you to sharpen your communication skills, as you will have to influence many people, and you will also have to handle objections that will require you to understand the principles of conflict management and negotiation techniques.

Inspiring individuals to perform to their best abilities has often resembled going on the journey to find the Holy Grail. For sure, you can get people to do their work by all sorts of coercive means; however, that does not signify they have the willingness or the competence to be successful.

7.3 Responding to Personal Drivers

Which brings us to the topic of motivation.

Great books, articles and treatises have been written on the subject of motivation, and I am extremely thankful to have read a number of these. These works expand on behavior and personality traits, some of which were even recognized in the times of antiquity—quoting Ezekiel, circa 590 BCE, *"strong as an ox, lethal as a lion, perceptive as an eagle"*, will give you an idea of how long behavior has been categorized.

However, behavior is *what* we do. It remains to be determined *why* we do it. And that leads us into a complex arrangement of psychological and physical drivers. I prefer to leave the exploration of these reasons to more competent individuals who have observed and derived the core reasons why people do what they do. The classic works on this subject are Abraham Maslow's *Motivation and Personality* and *Maslow on Management* (originally *Eupsychian Management: A Journal*) and Frederick Herzberg's *The Motivation to Work,* and there are many others.

So, what drives individuals to do what they do? Do people go to the workplace just for the money? Obviously, all of us will receive income from our work, so I am taking that off the table for now.

In the domain of project management, you are responsible for the results achieved by other people—principally, your team members, especially those who form the core of the team. In most cases, team members, when chosen appropriately, will have the necessary competence and skills to perform what is assigned to them.

For the most part, core team members and those who are closely peripherical to the core would be what are called *knowledge workers,* which implies that they have gone through formal higher-level education. They would have the intellectual capacity and, in the majority of cases, the skills to perform the project tasks that are assigned to them. In fact, I contend that all employees have the intellectual capacity to perform in the organization, and there is no demarcation between jobs categorized as white or blue collar (Wow, what a 19th-century expression!).

So, what makes some people excel at what they do and others underperform? Well, it depends first on why they *want* to do it and second if they actually *can* do it.

Let's get back to motivational theory and the drivers of motivation—why we do things. Once an individual can satisfy their "lower" physiological and safety needs, they release time and space to pursue "upper" needs (thank you again, Mr Maslow)—the psychological needs we all want to satisfy.

These upper needs are addressed according to each individual's priorities, and they would begin with wanting to be being part of a social group, as, I have mentioned above, we humans are communitarian. This will present opportunities for team-based work and interaction/networking with co-workers. Satisfying these needs—and each of us has different levels—liberates you to seek challenging work, giving you the freedom to make decisions and pursue opportunities for advancement, recognition and awards.

Eventually, when you can satisfy to an acceptable level this first set of upper needs, you can pursue the pinnacle of needs with enthusiasm to strive for growth, personal development and engaging in creative work.

The challenge for each of us is how to balance satisfying all these needs in the space of a day, and especially, as is my "propos", the famous minimum of nine hours a day in the workplace.

And that brings us back to what you as a project manager must do, starting with one simple negation:

> You do not motivate people—you must create an environment in which people find motivation.

You are the catalyst in creating such an environment, and you must focus hard on establishing it, sustaining it and making it blossom.

7.4 Inclusion and Unleashing Performance

Your first step is inclusion of individuals to be part of the group or structure, as people will, for the vast majority, satisfy their motivational driver of belonging. It is not just setting up a team, having a group meeting and moving on. It is consistently presenting and informing team members on the project, its status and its intricacies; identifying roles and responsibilities; engaging in bilateral sharing; listening and being empathetic to situations that individuals may be experiencing. It is creating an environment where people participate in and feel part of the project.

Inclusion in the group gives the individual an increased desire to participate and unleashes creativity and innovation in the pursuit of a superior performance that improves the group's achievements. Lack of inclusion reduces the individual's involvement, and such alienation reduces the individual's performance to no more than is required, whilst harboring a great sense of frustration.

Obviously the most involved will be your team members. However, you must extend this to all stakeholders, especially the key members who are supportive of the project. I suggest you also extend the dynamics to principal external providers, because they are part of the project group. Failing to generate such an ambiance will be sorely felt when you encounter discord and underperformance during the project execution. The conclusion is simple:

> People do not like to be excluded from a social group
> to which they consider they belong.

The teamwork environment you create will facilitate the distribution of activities and tasks, as team members will have already understood and established where and how they fit within the project. You will call upon classic delegation principles for the allocation of work—a great deal has been written on the subject; I invite you to explore the writings of authors such as Peter Drucker.

Delegation is not a one-way street, where you just hand out work to be done. The scope of the work has to be defined and analyzed to understand its nature and the types of resource profiles that would be required. You must map what needs to be done to those individuals who have or come close to the competence and skills required—and this is no mean task. You cannot assume that the person to whom the work is to be delegated has all the competence and the supporting resources to perform the work and complete it within the constraints of the project.

Discussion must ensue between you as the delegator and the delegate and agreement reached as to what needs to be produced and how that will be supported, verified and recognized. When competence or skills are lower than required, you must assess how you can assist the other person or, if necessary, where assistance can be found. Do not plough on regardless, as you will be committing the person to failure, and you will be the only person to blame when it comes to that.

When the delegated work is challenging and closely matches the team members' abilities, and you have given them the authority for decision-making with regards to the work delegated, you will satisfy and even increase those those individuals' motivational driver.

> People will out-perform when the work delegated
> fits their ability and aspirations.

7.5 Humility and Project Humanitarianism

When managing a project, you must put aside the belief that you are the most important person in the group. Otherwise, this will lead you to centralize decision-making, and you will alienate team members, who in many cases are more important to the project, as they may have hard and soft competences and skills that you do not possess.

Humility is in order, and your personal motivational drivers are satisfied by ensuring that the environment you have created, by being the sole catalyst, has enabled all team members to perform to the best of their abilities and accomplish the project's results successfully.

Beyond team members, you will coordinate and collaborate with stakeholders from the organization's functional departments. You will apply the same principles as those you deploy with the team; however, you will call upon additional communication skills to establish buy-in and commitment from these participants. You must position yourself to be aware of political agendas whilst not playing a part in them. Your job is the organization's project and should not be subject to petty infights. Remember, you will never satisfy everyone, so in many cases you need to be humble and non-confrontational.

Managing people is a life changer which challenges your education, presumptions and culture and pits these against the harsh realities of functioning in a social group. Motivation makes all the difference—challenge your assumptions about human nature.

A great number of errors are made by project managers who lack the knowledge, competence, skills and human behavior required for the job. Or they just prefer to revel in a position of superiority and make things worse by acting as mini authoritarians or petty dictators.

To bring a holistic approach to sanity, it is essential that the organization provide for continuous training courses and refreshers on the spectrum of human skills that project managers must possess. Only then can the organization envisage achieving its goals and objectives supported by a willing, able and committed workforce—*the people.*

> The most important sanity to promote is the human and people side of project management.

7.6 Chapter 7 Summary

Central to working and engaging with all participants of a project are the so-called *soft-skills* (whoever invented this terminology should be sent far away to outer space).

Working in groups and in collaboration has been the hallmark of humanity; a project is performed by a group, and failure is assured when there is lack of purpose and cohesion for any given group.

All groups need catalysts, and these may be any member of the group. However, the central and main catalyst is the project manager, as that person is regarded as the temporary head of the venture. All participants will initially follow humanity's protocol and abide by the given spontaneous hierarchy of the project. The onus is on the project manager to demonstrate skills that can validate this hierarchy.

Over millennia, social groups have been managed by terror, autocracy, dictatorship and other forceful means, and many have succeeded. Unfortunately, this is still the case in the 21st century.

However, a project manager cannot use the above playbook and assume success will be assured. Furthermore, the project manager is on a temporary assignment, working with individuals who are not under a hierarchical or direct reporting mechanism that can be enforced. Those who have succeeded in managing in the many different group situations throughout history have not relied on hierarchical power (unless forced to).

The vast majority of people engaged in projects earn their living by working in an organization. They spend the legal number of days and hours, and more, performing the assigned tasks according to their abilities. Individuals want to enjoy their work experience and be recognized for their efforts.

The project manager's role is as a coordinator, communicator, orchestrator, animator and, most of all, an interpreter of purpose and goals. Leadership skills have to be honed, and the understanding of motivational drivers is a central part of that role. Interpersonal communications must be fine-tuned to all members of the project group. This extends to communicating to the large group of participants.

Without entering the everlasting discussion on whether a project manager may possess such skills innately, suffice it to say that all these skills can be learnt or perfected given exposure and experience. And that is where organizational management must conduct extensive soft-skills training to all current and prospective project managers.

Index

Printed in the United States
by Baker & Taylor Publisher Services